FLOW-SCHEMES IN ORGANIC CHEMISTRY

Colin McCarty B.Tech., Ph.D.

Chairman of the Science Faculty,
Carisbrooke High School,
Newport, Isle of Wight

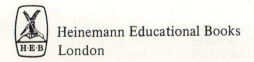

Heinemann Educational Books
London

Heinemann Educational Books Ltd
22 Bedford Square, London WC1B 8HH
LONDON EDINBURGH MELBOURNE AUCKLAND
HONG KONG SINGAPORE KUALA LUMPUR NEW DELHI
IBADAN NAIROBI JOHANNESBURG
EXETER (NH) KINGSTON PORT OF SPAIN

ISBN 0 435 65575 2

Set in IBM Press Roman by Tecmedia Ltd
Printed and bound in Great Britain by
Spottiswoode Ballantyne Ltd, Colchester and London

Preface

The format of the flow-schemes used in this book is a departure from the standard convention for writing an organic reaction sequence, in that it highlights the reagent rather than the starting material and product. This style has been chosen to align the logic of organic synthesis to the logic diagrams that are taught in modern mathematics syllabuses: namely, the operation is 'in the box'. Thus the starting compound and the product of the reaction are, respectively, before and after the operation.

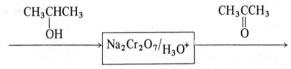

This book does not attempt to cover the reactions or preparations in the same depth as a textbook; its purpose is rather to give the student an opportunity to utilize his knowledge by applying it to solve and complete the flow-schemes. In so doing it is hoped that the student will gain expertise in piecing reactions and preparations together into logical sequences, which will enable him to understand something of the art of devising an organic synthesis; and also make him more aware of the relationships between the various functional groups in organic chemistry.

The book is divided into five sections

It is recommended that the simple two-step syntheses are tackled and understood before the flow-schemes are attempted.

To aid solving the flow-schemes there are two indexes which refer back to the synopses:

1 reactions of functional groups and

2 reagents.

The synopses are grouped into preparations of functional groups, e.g. alcohols, ketones, etc., for ease of reference if a preparation is sought.

It should be emphasized that this book is intended to provide an interesting and stimulating way of checking, testing, and revising in organic chemistry and is not for those first learning organic chemistry. It has been designed essentially

iii

for 'A' level students, and those taking O.N.C. and H.N.C. courses. The synopses cover the great majority of organic reactions listed in the new J.M.B. and London syllabuses, together with the contents of the Nuffield 'A' level books.

With only a few exceptions, the recommendations of the *Report on Chemical Nomenclature, Symbols, and Terminology for use in School Science,* prepared by a working party of the Education (Research) Committee of the Association for Science Education, have been adhered to.

I should like to express my thanks to Miles Hedges, John Mullett, Andy Marshall, and Stephen Wise for their help in testing the flow-schemes. The original ideas for these flow-schemes was suggested to me by Michael Cane of Bloxham School. I am particularly indebted to Robin Hillman, of Bloxham, and Martyn Berry, of Chislehurst and Sidcup Grammar School, for their constructive criticism of the synopses and help in testing the flow-schemes. I should also like to thank Dr K. G. Mason and Dr H. Heaney, of Loughborough University of Technology, for their valuable suggestions and interest shown in this book. Finally I should like to thank Miles Hedges for reading the manuscript.

1976   C. McC.

Contents

Synopses of Preparations

Preparation of halogenoalkanes

Substitution of an alcohol

$$\overset{|}{\underset{|}{-}}\text{C--OH} \longrightarrow \overset{|}{\underset{|}{-}}\text{C--X} \qquad \text{(where X = Cl, Br or I)}$$

A number of reagents can be used and are tabulated below.

To drive this equilibrium reaction to the right, dehydrating agents may be added to remove the by-product, water, and so increase the yield of halogeno-alkane.

Elimination of water is a competing side reaction and, in tertiary alcohols in particular, significant amounts of alkene are produced.

Product	Reagents		
Chloroalkane	PCl_3 or PCl_5	$SOCl_2$	$HCl/_{ZnCl_2}$
Bromoalkane	PBr_3	$NaBr/_{H_2SO_4}$	HBr
Iodoalkane	PI_3 (red $P/_{I_2}$)	HI	

Decarboxylation of a silver carboxylate

$$\overset{|}{\underset{|}{-}}\text{C--CO}_2^-\text{Ag}^+ \longrightarrow \overset{|}{\underset{|}{-}}\text{C--Br}$$

When a silver salt of a carboxylic acid, prepared by treating the acid with moist silver oxide, is heated under reflux with bromine in tetrachloromethane, a bromoalkane is formed and carbon dioxide given off. This is a useful way of descending a homologous series by one carbon atom.

$$\begin{matrix} CH_3 \\ CH_3 \\ CH_3 \end{matrix}\!\!\!\succ\!\! CCH_2CO_2^-Ag^+ \longrightarrow \boxed{Br_2/_{CCl_4}} \longrightarrow \begin{matrix} CH_3 \\ CH_3 \\ CH_3 \end{matrix}\!\!\!\succ\!\! CCH_2Br + CO_2$$

yield 53%

Addition of hydrogen halide to an alkene

$$\ce{C=C} \longrightarrow \ce{H-C-C-X} \quad \text{(where X = Cl, Br or I)}$$

Hydrogen halides readily add to double bonds to form halogenoalkanes. Two products are possible if the double bond is not symmetrically substituted. The major product is the halogenoalkane with the halogen bonded to the carbon atom which has the most carbon atoms (or electron doning groups) attached to it (Markownicoff's rule).

$$\underset{CH_3}{\overset{CH_3}{>}}C=CH_2 \longrightarrow \boxed{HCl\,(g)} \longrightarrow CH_3-\underset{CH_3}{\overset{CH_3}{\underset{|}{\overset{|}{C}}}}-Cl$$

yield 100%

Addition of halogen to an alkene

$$\ce{C=C} \longrightarrow \overset{X}{\underset{X}{-C-C-}} \quad \text{(where X = Br or Cl)}$$

Halogens add readily, and in many cases quantitatively, to double bonds. A convenient solvent for such reactions is trichloromethane.

$$CH_2=CH_2 \longrightarrow \boxed{Br_2} \longrightarrow CH_2BrCH_2Br$$

yield 100%

Substitution of an alkane by a halogen

$$-\overset{|}{\underset{|}{C}}-H \longrightarrow -\overset{|}{\underset{|}{C}}-X \quad \text{(where X = Br or Cl)}$$

Halogen radicals, generated by ultra-violet light on the gaseous halogen, will substitute hydrogen atoms on an alkane chain with little discrimination. Benzene side chains, however, are specifically activated in the positions adjacent to the benzene ring.

$$CH_3-\langle\bigcirc\rangle-CH_2CH_3 \longrightarrow \boxed{Cl_2/light} \longrightarrow CCl_3-\langle\bigcirc\rangle-\underset{Cl}{\overset{Cl}{\underset{|}{\overset{|}{C}}}}CH_3$$

Substitution of an arene

$$\bigcirc \longrightarrow \bigcirc-X \quad \text{(where X = Br or Cl)}$$

When an arene is heated with halogen in the presence of a catalyst such as iron(III) chloride or the respective aluminium halide, good yields of the halogenoarene are obtained.

$$\underset{\displaystyle \bigcirc}{\overset{\displaystyle NO_2}{|}} \longrightarrow \boxed{Br_2/AlBr_3} \longrightarrow \underset{\displaystyle \bigcirc}{\overset{\displaystyle NO_2}{|}} Br$$

yield 96%

Substitution of an arenediazonium ion

$$\bigcirc-N_2^+ \longrightarrow \bigcirc-X \quad \text{(where X = Cl, Br or I)}$$

Warming the arenediazonium salt with hydrogen halide, in the presence of the respective copper(I) halide, gives good yields of the halogenoarene.

$$\underset{\displaystyle \bigcirc}{\overset{\displaystyle CH_3}{|}}-N_2^+Cl^- \longrightarrow \boxed{HCl/CuCl} \longrightarrow \underset{\displaystyle \bigcirc}{\overset{\displaystyle CH_3}{|}}-Cl$$

yield 76%

Substitution at a 2-carbon atom of a carbonyl – containing compound

$$\underset{-C-C}{\overset{H \quad \quad O}{\diagdown}} \longrightarrow \underset{-C-C}{\overset{X \quad \quad O}{\diagdown}} \quad \text{(where X = Br or Cl)}$$

Hydrogen atoms bonded to the carbon atom adjacent to a carbonyl function are readily replaced by a halogen. In this reaction aldehydes and ketones are catalyzed by both acid and base; acyl halides and carboxylic acids by phosphorus.

$$CH_3CH_2CO_2H \longrightarrow \boxed{Br_2/P} \longrightarrow \underset{\quad\quad\quad Br}{\overset{}{CH_3CHCO_2H}}$$

Preparation of alcohols

Reduction of a carboxylic acid

$$-C\overset{O}{\underset{OH}{\big\langle}} \longrightarrow \overset{H}{\underset{H}{\big\rangle}}C-OH$$

A convenient laboratory procedure is to add excess lithium aluminium hydride to a solution of the carboxylic acid in dry ethoxyethane. Unreacted lithium aluminium hydride is destroyed with water and, provided the amount of water is kept to a minimum, the alcohol may be recovered from the organic phase.

A primary alcohol is always formed.

$$\underset{CH_3}{\overset{CH_3}{\big\rangle}}CHCO_2H \longrightarrow \boxed{LiAlH_4} \longrightarrow \underset{CH_3}{\overset{CH_3}{\big\rangle}}CHCH_2OH$$

yield 80%

Reduction of an ester

$$-C\overset{O}{\underset{OR}{\big\langle}} \longrightarrow \overset{H}{\underset{H}{\big\rangle}}C-OH + ROH$$

Conditions similar to those for the reduction of carboxylic acids are used, however, two alcohols are produced and separation of the products may be more complicated.

For esters, but not carboxylic acids, catalytic reduction with hydrogen, or reduction by sodium in ethanol may also be used.

Reduction of an aldehyde

$$-C\overset{O}{\underset{H}{\big\langle}} \longrightarrow \overset{H}{\underset{H}{\big\rangle}}C-OH$$

Lithium aluminium hydride or catalytic hydrogenation will reduce aldehydes to primary alcohols. A milder reagent that is frequently used is sodium borohydride dissolved in aqueous ethanol.

The advantage of lithium aluminium hydride and sodium borohydride is that they are selective reducing agents; carbon-carbon double bonds present in the molecule are not usually reduced.

$$CH_3CH=CHCHO \longrightarrow \boxed{LiAlH_4} \longrightarrow CH_3CH=CHCH_2OH$$

Reduction of a ketone

$$\overset{\diagdown}{\underset{\diagup}{}}C=O \longrightarrow -\overset{H}{\underset{\diagup}{C}}-OH$$

Conditions similar to those for the reduction of aldehydes are used; however a secondary alcohol is produced.

$$CH_3CH_2\underset{\substack{\| \\ O}}{C}CH_3 \longrightarrow \boxed{LiAlH_4} \longrightarrow CH_3CH_2\underset{\substack{| \\ OH}}{C}HCH_3$$

yield 90%

Addition to an alkene

$$\overset{\diagdown}{\underset{\diagup}{}}C=C\overset{\diagup}{\underset{\diagdown}{}} \longrightarrow H-\overset{|}{C}-\overset{|}{C}-OH$$

Concentrated sulphuric acid is added carefully to the alkene, to form an alkyl hydrogensulphate. This, when poured into cold water, is hydrolysed to give the alcohol. The alcohol group will be attached to the carbon atom which has the most carbon atoms bonded to it (Markownicoff's rule).

$$\underset{H}{\overset{CH_3}{}}\diagdown C=C \underset{CH_3}{\overset{CH_3}{}} \longrightarrow \boxed{\begin{array}{c} H_2SO_4 \\ \text{then } H_2O \end{array}} \longrightarrow CH_3CH_2\underset{\substack{| \\ OH}}{\overset{\substack{CH_3 \\ |}}{C}}CH_3$$

yield 74%

Substitution of a halogenoalkane

$$-\overset{|}{C}-X \longrightarrow -\overset{|}{C}-OH \quad \text{(where X = Cl, Br or I)}$$

Heating a halogenoalkane with sodium hydroxide in ethanol will produce some of the desired product, but elimination of hydrogen halide readily occurs under these conditions to form an alkene. Rearrangement of the carbonium ion intermediate, formed when tertiary and many secondary halogenoalkanes are treated in the above manner, also cuts down the yield of the required alcohol. Refluxing the halogenoalkane with an aqueous suspension of calcium carbonate, $CaCO_3$, a weak base, may be used to cut down the amount of alkene by-product and increase the yield of alcohol.

Oxidation of an alkene to a diol

$$\text{C=C} \longrightarrow \text{HO-C-C-OH}$$

Neutral potassium manganate(VII) gives variable yields of 1,2-diol. Under basic conditions further oxidation may occur to form ketones or carboxylic acids, by cleavage of the carbon-carbon bond.

$$\text{(cyclohexene)} \longrightarrow \boxed{KMnO_4/H_2O} \longrightarrow \text{(cyclohexane-1,2-diol, OH, OH)}$$

Hydrolysis of an epoxide

$$\underset{O}{\text{C-C}} \longrightarrow \underset{HO}{\text{-C-C-}}^{OH}$$

The epoxide, when treated with aqueous acid, is smoothly converted into the diol.

When an epoxide is warmed with alcoholic acid the 2-alkoxyalcohol is formed. If the epoxide is not symmetrically substituted, two products can be formed. The major product is the 'least substituted' alcohol.

$$CH_3CH-CH_2 \longrightarrow \boxed{C_2H_5OH/H_3O^+} \longrightarrow CH_3CHCH_2OH$$

$$\underset{OC_2H_5}{|}$$

major product

$$\longrightarrow CH_3CHCH_2OC_2H_5$$

$$\underset{OH}{|}$$

minor product

Hydrolysis of an ester

$$R-C\underset{O-C-}{\overset{O}{\diagup}} \longrightarrow -\overset{\cdot}{C}OH + RCO_2^- Na^+$$

Base hydrolysis of an ester will give good yields of the alcohol together with the salt of the carboxylic acid. The alcohol may be separated by extraction into ethoxyethane.

Substitution of a diazonium salt

$$\langle\bigcirc\rangle-N_2^+ \longrightarrow \langle\bigcirc\rangle-OH$$

Arenediazonium salts are prepared by treating the corresponding amine with a cold solution of sodium nitrate(III) (sodium nitrite) in dilute hydrochloric acid (nitrous acid). The diazonium salt is fairly stable in solution below 280 K. Above room temperature hydrolysis occurs to form the phenol, and this may be used preparatively.

$$\langle\bigcirc\rangle-NH_2 \rightarrow \boxed{\begin{array}{c}NaNO_2/HCl \\ 280\ K\end{array}} \rightarrow \langle\bigcirc\rangle-N_2^+Cl^- \rightarrow \boxed{\begin{array}{c}H_2O \\ 320\ K\end{array}} \rightarrow \langle\bigcirc\rangle-OH$$

Substitution of a sulphonic acid

$$\langle\bigcirc\rangle-SO_3H \longrightarrow \langle\bigcirc\rangle-OH$$

The arenesulphonic acid is fused with solid sodium hydroxide to form the sodium salt of the phenol. Neutralization with aqueous acid gives the phenol which can be extracted into an organic solvent.

$$\langle\bigcirc\rangle-SO_3H \rightarrow \boxed{\begin{array}{c}NaOH/ \\ heat\end{array}} \rightarrow \langle\bigcirc\rangle-O^-Na^+ \rightarrow \boxed{H_3O^+} \rightarrow \langle\bigcirc\rangle-OH$$

Preparation of Alkenes

Elimination from an alcohol

$$\underset{OH}{\overset{H}{\underset{|}{-C-C-}}} \longrightarrow \overset{\diagdown}{\underset{\diagup}{C}}=\overset{\diagup}{\underset{\diagdown}{C}}$$

Dehydration of an alcohol will give the alkene. A number of reagents are commonly used including: phosphorus pentoxide, P_2O_5, phosphoric acid, H_3PO_4, and sulphuric acid, H_2SO_4. Lower yields are sometimes obtained with sulphuric acid as ethers may be formed by reaction of the alcohol with the intermediate sulphate. Generally where two products may be formed the 'most substituted' double bond is the predominant product.

yield 85%

Elimination from a halogenoalkane

$$\underset{X}{\overset{H}{\underset{|}{-C-C-}}} \longrightarrow \overset{\diagdown}{\underset{\diagup}{C}}=\overset{\diagup}{\underset{\diagdown}{C}} \qquad \text{(where X = Cl, Br or I)}$$

Heating a halogenoalkane under reflux with an ethanolic solution of potassium hydroxide gives variable yields of alkene. Substitution to form an alcohol is a competitive reaction. Tertiary halogenoalkanes undergo elimination most readily and primary least readily.

Where two products may be formed the 'most substituted' double bond is the major product.

86% of the yield

14% of the yield

Preparation of Ketones

Oxidation of a secondary alcohol

$$\underset{\underset{H}{\overset{H}{|}}}{-\!\!\overset{\displaystyle\,}{C}\!-OH} \longrightarrow \;\; \overset{\diagdown}{\underset{\diagup}{C}}\!=\!O$$

The secondary alcohol is heated with acidic sodium dichromate(VI). Ethanoic acid is often used as the solvent. The ketone may either be distilled from the reaction mixture or, if it is not sufficiently volatile, the solution may be neutralized with sodium carbonate and the ketone extracted into an organic solvent.

$$\underset{CH_3}{\overset{CH_3}{\diagdown}}\!CHOH \longrightarrow \boxed{Na_2Cr_2O_7/H_3O^+} \longrightarrow \underset{CH_3}{\overset{CH_3}{\diagdown}}\!C\!=\!O$$

yield 85%

Oxidative cleavage of a double bond

$$\overset{\diagdown}{\underset{\diagup}{C}}\!=\!\overset{\diagup}{\underset{\diagdown}{C}} \longrightarrow \overset{\diagdown}{\underset{\diagup}{C}}\!=\!O + O\!=\!\overset{\diagup}{\underset{\diagdown}{C}}$$

Ozone-enriched oxygen is bubbled through the solution containing the compound, which is cooled by solid carbon dioxide to about 200 K. The ozonide intermediate so formed is warmed with zinc in aqueous ethanoic acid and decomposes to form the ketones. Two ketones will only be formed when no hydrogen atoms are attached to the carbon-carbon double bond.

$$\underset{CH_3}{\overset{CH_3}{\diagdown}}\!C\!=\!C\underset{C_3H_7}{\overset{C_2H_5}{\diagup}} \longrightarrow \boxed{\begin{array}{c}O_3 \text{ then}\\ Zn, H_3O^+\end{array}} \longrightarrow \underset{CH_3}{\overset{CH_3}{\diagdown}}\!C\!=\!O + O\!=\!C\underset{C_3H_7}{\overset{C_2H_5}{\diagup}}$$

Substitution of an arene by an acyl chloride

$$\bigcirc \longrightarrow \bigcirc\!\!-\!\!C\!\!\overset{\displaystyle O}{\underset{R}{\diagdown}}$$ (where R = alkyl or aryl)

Heating an acyl chloride and an arene together in the presence of anhydrous aluminium chloride gives good yields of the aryl ketone. (This is a type of Friedel-Crafts reaction.)

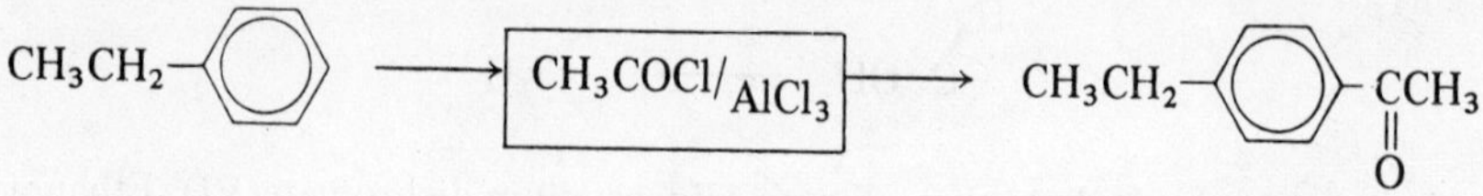

yield 98%

Preparation of Aldehydes

Reduction of an acyl chloride

Reduction, using hydrogen and a partially poisoned palladium catalyst, gives moderate to good yields of aldehyde. The ease with which aldehydes are reduced to alcohols means care must be taken to ensure ideal conditions are present for a particular compound. (This is known as the Rosenmund reaction.)

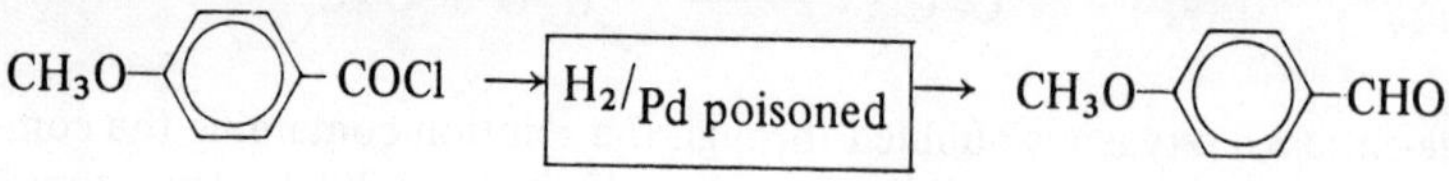

yield 81%

Oxidation of a primary alcohol

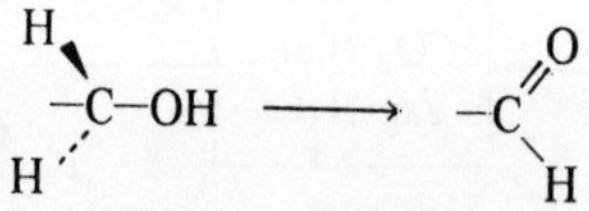

Heating a primary alcohol with an acidified sodium dichromate(VI) solution will form the aldehyde, however it must be distilled out as it is formed otherwise further oxidation to the carboxylic acid will occur. Thus the use of this reaction is limited to preparing only those aldehydes with low boiling points.

Oxidative cleavage of a double bond

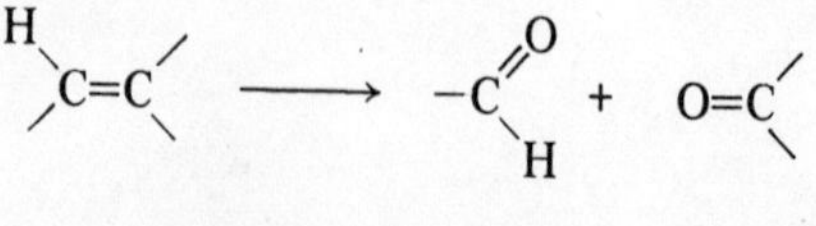

Ozone-enriched oxygen is bubbled through a solution containing the compound, which has been cooled to below 200 K. The ozonide intermediate so formed is decomposed by zinc and dilute aqueous ethanoic acid to form a mixture of carbonyl-containing products. If a peroxoacid is used to decompose the ozonide a carboxylic acid rather than an aldehyde is produced.

$$\text{(cyclohexene)} \longrightarrow \boxed{\begin{array}{c} O_3 \text{ then} \\ Zn, H_3O^+ \end{array}} \longrightarrow \begin{array}{c} \text{—CHO} \\ \text{—CHO} \end{array}$$

yield 60%

Preparation of Carboxylic Acids

Oxidation of a primary alcohol

$$\begin{array}{c} H \\ -C-OH \\ H \end{array} \longrightarrow -C\!\!\begin{array}{c} \nearrow O \\ \searrow OH \end{array}$$

The primary alcohol is heated under reflux with an acidic solution of sodium dichromate(VI). Ethanoic acid is often a convenient solvent for this reaction. With complex molecules milder reagents are used, such as alkaline potassium manganate(VII). In this case the potassium salt of the carboxylic acid is formed.

$$C_5H_{11}CH_2OH \longrightarrow \boxed{\begin{array}{c} KMnO_4/OH^- \\ \text{then } H_3O^+ \end{array}} \longrightarrow C_5H_{11}CO_2H$$

yield 75%

Oxidation of an aldehyde

$$-C\!\!\begin{array}{c} \nearrow O \\ \searrow H \end{array} \longrightarrow -C\!\!\begin{array}{c} \nearrow O \\ \searrow OH \end{array}$$

Similar, or milder, conditions to those used for the oxidation of alcohols will convert aldehydes to carboxylic acids. Moist silver oxide, Ag_2O, or ammoniacal silver nitrate, $Ag(NH_3)^+$, (Tollen's reagent) will oxidize the aldehyde and the silver(I) ions will be reduced to metallic silver. Fehling's reagent also oxidizes aldehydes; the blue copper(II) complex is reduced to copper(I) oxide, a brown precipitate.

$$CH_3CH=CHCHO \longrightarrow \boxed{Ag_2O/H_2O} \longrightarrow CH_3CH=CHCO_2H$$

yield 60%

Hydrolysis of an amide

$$-C\overset{O}{\underset{NH_2}{\diagdown}} \longrightarrow -C\overset{O}{\underset{OH}{\diagdown}}$$

Amides may be hydrolysed by either acids or bases. As the reactions are relatively slow, the usual procedure is to heat the reaction mixture under reflux.

$$CH_3CONH_2 \longrightarrow \boxed{H_3O^+, 370\,K} \longrightarrow CH_3CO_2H$$

yield 60%

Hydrolysis of a nitrile

$$-C\equiv N \longrightarrow -C\overset{O}{\underset{NH_2}{\diagdown}} \longrightarrow -C\overset{O}{\underset{OH}{\diagdown}}$$

Nitriles are readily hydrolysed to amides which are further hydrolysed to carboxylic acids by more severe conditions, namely heating under reflux.

2-hydroxyacids and 2-aminoacids can be prepared respectively by hydrolysing the 2-hydroxyalkanonitrile (cyanohydrin) and 2-aminoalkanonitrile.

$$\text{⬡—CHO} \rightarrow \boxed{\text{HCN/}_{\text{base}}} \rightarrow \text{⬡—}\underset{OH}{\overset{}{CHCN}} \rightarrow \boxed{\substack{H_3O^+, \\ 370\,K}} \rightarrow \text{⬡—}\underset{OH}{\overset{}{CHCO_2H}}$$

yield 86%   yield 60%

$$CH_3CHO \rightarrow \boxed{NH_4^+CN^-} \rightarrow CH_3\underset{NH_2}{CHCN} \rightarrow \boxed{\substack{H_3O^+, \\ 370\,K}} \rightarrow CH_3\underset{NH_2}{CHCO_2H}$$

yield 60%

Cleavage of a methylketone

$$\underset{}{\overset{CH_3}{\diagdown}}C{=}O \longrightarrow -C\overset{O}{\underset{OH}{\diagdown}}$$

Methyl ketones react readily with iodine in sodium hydroxide solution to form triiodomethane and the sodium salt of the carboxylic acid. Triiodomethane is a yellow, water-insoluble compound and this is therefore a good reaction to show the presence of a methyl group adjacent to a carbonyl group.

Chlorine or bromine in sodium hydroxide solution may also be used and will form the corresponding trihalogenomethane.

$$\underset{CH_3}{\overset{CH_3}{>}}CHCCH_3 \quad \xrightarrow{\boxed{\begin{array}{c} Br_2/OH^- \\ \text{then } H_3O^+ \end{array}}} \quad \underset{CH_3}{\overset{CH_3}{>}}CHCO_2H \;+\; CHBr_3$$

yield 85%

Hydrolysis of an ester

$$-\overset{O}{\underset{OR}{\overset{\|}{C}}} \quad \longrightarrow \quad -\overset{O}{\underset{OH}{\overset{\|}{C}}} \;+\; ROH$$

Boiling an ester with sodium hydroxide will form the sodium salt of the carboxylic acid and an alcohol. Neutralization of the mixture with dilute mineral acid gives the carboxylic acid. Acid hydrolysis can also be used but yields are often poor as the reaction is an equilibrium one in which the position of equilibrium may favour the ester.

$$\begin{array}{l} CH_2O_2CC_{17}H_{35} \\ CHO_2CC_{17}H_{35} \\ CH_2O_2CC_{17}H_{35} \end{array} \to \boxed{NaOH} \to \begin{array}{l} CH_2OH \\ CHOH \\ CH_2OH \end{array} + \underset{\text{'soap'}}{C_{17}H_{35}CO_2^-Na^+} \to \boxed{H_3O^+} \to C_{17}H_{35}CO_2H$$

Hydrolysis of an acyl chloride

$$-\overset{O}{\underset{Cl}{\overset{\|}{C}}} \quad \longrightarrow \quad -\overset{O}{\underset{OH}{\overset{\|}{C}}}$$

Acyl chlorides are rapidly hydrolysed by water to form carboxylic acids. As acyl chlorides are prepared from carboxylic acids this can hardly be called a preparative procedure.

Hydrolysis of an anhydride

$$\begin{array}{c} R'-\overset{O}{\overset{\|}{C}} \\ \quad\;\; O \\ R''-\overset{}{\underset{O}{\overset{}{C}}} \end{array} \longrightarrow R'-\overset{O}{\underset{OH}{\overset{\|}{C}}} \;+\; R''-\overset{O}{\underset{OH}{\overset{\|}{C}}}$$

With water, anhydrides hydrolyse to form carboxylic acids. If an alcohol is used instead of water, an ester and a carboxylic acid are produced. This latter reaction is more useful synthetically, as anhydrides are prepared from carboxylic acids.

Hydrolysis of a diazotized amide

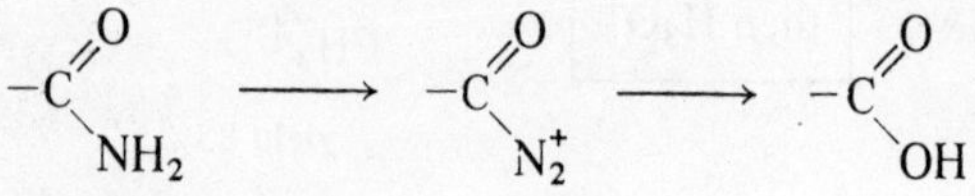

Adding sodium nitrate(III) to an amide dissolved in a mixture of ethanol and dilute aqueous sulphuric acid gives high yields of the carboxylic acid. This reaction is particularly valuable for use with amides which are in molecules that are sterically hindered to base or acid hydrolysis.

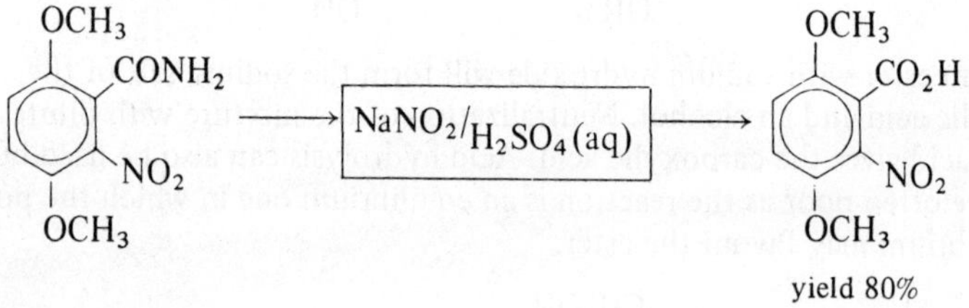

yield 80%

Preparation of Sulphonic Acids

Substitution of an arene

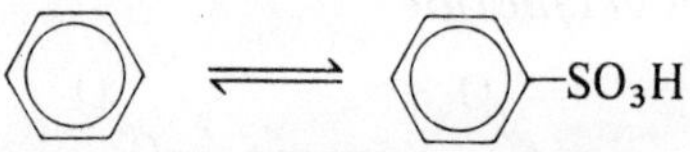

Heating an arene with a slight excess of fuming sulphuric acid will produce the sulphonic acid.

Preparation of Nitroarenes

Substitution of an arene

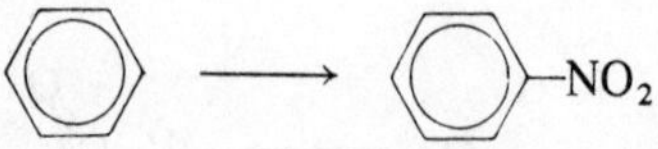

An arene may be nitrated by heating it with a mixture of concentrated sulphuric acid and concentrated nitric acid. More than one nitro group may be added using higher temperatures and fuming, rather than concentrated, acids.

Preparation of Acyl Chlorides

Replacement of —OH in a carboxylic acid

$$-C\overset{O}{\underset{OH}{\Big\langle}} \longrightarrow -C\overset{O}{\underset{Cl}{\Big\langle}}$$

Either phosphorus pentachloride, PCl_5, or sulphur dichloride oxide (thionyl chloride), $SOCl_2$, is added in small portions to a cooled, stirred solution of the carboxylic acid in a dry, inert solvent such as toluene. The acyl chloride is usually obtained by distilling it from the reaction mixture under reduced pressure.

$$\text{⬡}-CO_2H \longrightarrow \boxed{PCl_5} \longrightarrow \text{⬡}-COCl$$

Preparation of Anhydrides

Replacement of —H in a carboxylic acid

$$-C\overset{O}{\underset{OH}{\Big\langle}} \longrightarrow \begin{array}{c} -C\overset{O}{\diagdown} \\ \diagdown O \\ R-C\diagup \\ \diagdown O \end{array}$$

Dry solutions of a carboxylic acid and an acyl chloride are stirred together in the presence of an organic base such as pyridine. Alternatively, the sodium salt of the carboxylic acid and a dry solution of acyl chloride are stirred together.

$$CH_3CH_2CO_2H \longrightarrow \boxed{CH_3COCl/_{base}} \longrightarrow \begin{array}{c} CH_3CH_2C\overset{O}{\diagdown} \\ \diagdown O \\ CH_3C\diagup \\ \diagdown O \end{array}$$

Elimination of water from a 1,2-dicarboxylic acid

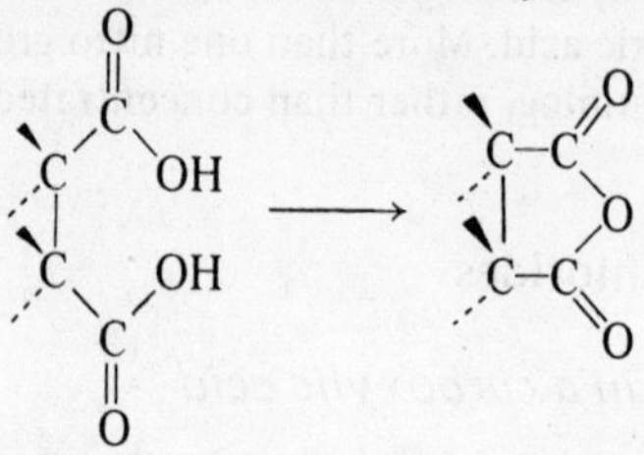

1,2-Dicarboxylic acids lose water when heated to about 500 K, to give variable yields of the anhydride.

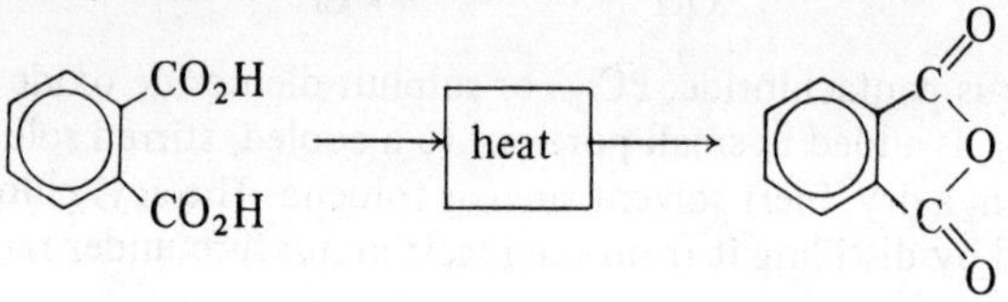

Preparation of Esters

Condensation of a carboxylic acid and an alcohol

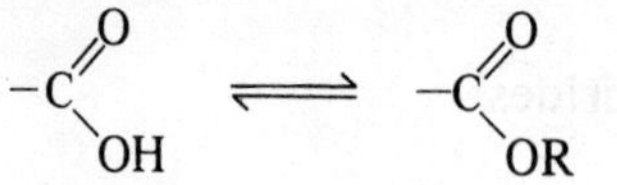

(R = alkyl or aryl group)

The establishment of equilibrium between carboxylic acid and alcohol, and ester and water, is catalysed by acid. Concentrated sulphuric acid may be used as it is also a dehydrating agent and so pulls the equilibrium to the right, thus increasing the yield of ester.

$$CH_3CO_2H \longrightarrow \boxed{C_2H_5OH/H_2SO_4} \longrightarrow CH_3CO_2C_2H_5$$

yield 80%

Condensation of an acyl chloride and an alcohol

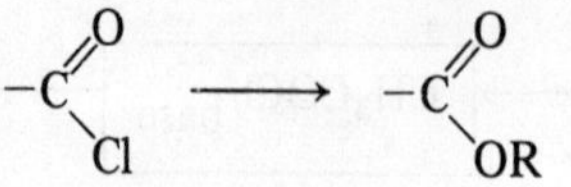

(R = alkyl or aryl group)

The acyl chloride and alcohol readily react together, particularly if a base is present to neutralize the hydrogen chloride which is eliminated.

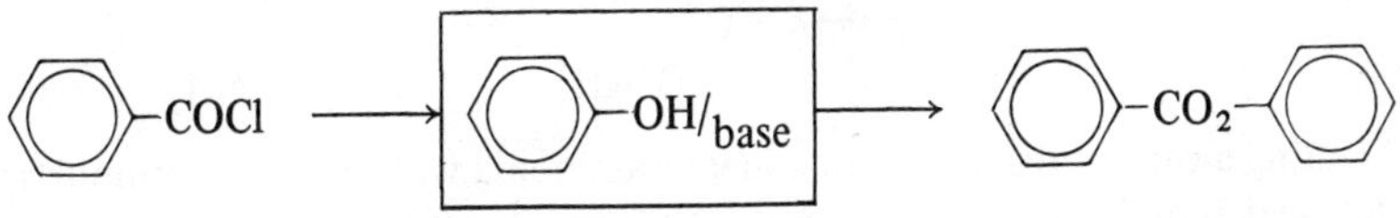

yield 75%

Condensation of an anhydride and an alcohol

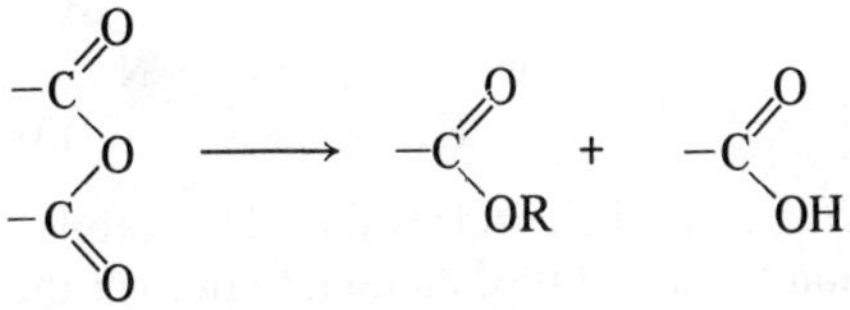

The anhydride and alcohol are heated under reflux for about 10 minutes, cooled, and then the mixture may be neutralized with sodium hydrogen carbonate solution. The ester can then be extracted into an organic solvent such as ethoxyethane.

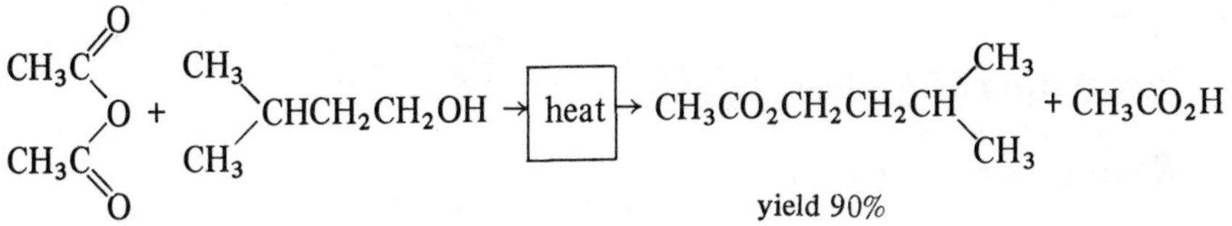

yield 90%

Preparation of Amides

Addition of water to a nitrile

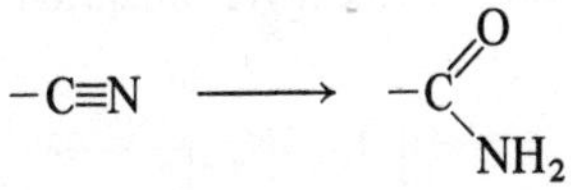

Gently heating a nitrile (below 320 K) with either aqueous acid or base will produce the amide. Higher temperatures must not be employed as the amide may be further hydrolysed to the carboxylic acid.

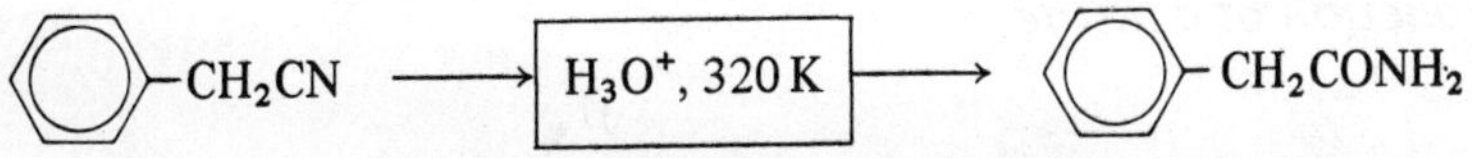

Heating the ammonium salt of a carboxylic acid

$$-C{\overset{O}{\underset{OH}{\big\langle}}} \longrightarrow -C{\overset{O}{\underset{O^- NH_4^+}{\big\langle}}} \longrightarrow -C{\overset{O}{\underset{NH_2}{\big\langle}}}$$

Treating a carboxylic acid with aqueous ammonia will form the ammonium salt, which on heating will lose water to form the amide.

Acetylation of an amine

$$-N{\overset{H}{\underset{H}{\big\langle}}} \longrightarrow -N{\overset{H}{\underset{OAc}{\big\langle}}} \qquad (Ac = acyl\ group)$$

An anhydride or an acyl chloride will readily acetylate an amine or ammonia. The acetylation of amines is often used to protect the amine function before undertaking a reaction on another part of the molecule, the amide can then be reduced to return the amine.

$$NH_3(aq) \longrightarrow \boxed{CH_3CH_2COCl} \longrightarrow CH_3CH_2CONH_2$$

Preparation of Amines

Reduction of an oxime

$$\underset{/}{\overset{\backslash}{C}}{=}N{\underset{OH}{\big\backslash}} \longrightarrow H{-}\underset{\cdot\cdot}{C}{-}NH_2$$

Excess lithium aluminium hydride is added to a solution of the oxime in dry ethoxyethane. Remaining unreacted lithium aluminium hydride is destroyed by the addition of water and the organic layer separated.

$$\overset{C_6H_5}{\underset{C_6H_5}{\big\rangle}}C{=}NOH \longrightarrow \boxed{LiAlH_4} \longrightarrow \overset{C_6H_5}{\underset{C_6H_5}{\big\rangle}}CHNH_2$$

yield 60%

Reduction of a nitrile

$$-C{\equiv}N \longrightarrow \overset{H}{\underset{H}{\underset{\cdot\cdot}{C}}}{-}NH_2$$

Using similar conditions to the reduction of an oxime, a nitrile may be reduced to form an amine.

A primary amine is always formed when a nitrile is reduced.

$$CH_3CH_2CH_2CN \longrightarrow \boxed{LiAlH_4} \longrightarrow CH_3CH_2CH_2CH_2NH_2$$

yield 57%

Reduction of an amide

$$-C\overset{O}{\underset{NH_2}{\diagup}} \longrightarrow -\overset{H}{\underset{H}{C}}-NH_2$$

Using similar conditions to the reduction of an oxime, an amide may be reduced to form an amine.

A primary amine is only formed when the amide function is not substituted.

$$CH_3C\overset{C_6H_5}{\underset{\underset{O}{\|}}{N}}\diagdown CH_3 \longrightarrow \boxed{LiAlH_4} \longrightarrow CH_3CH_2N\overset{C_6H_5}{\diagdown CH_3}$$

yield 91%

Substitution of a halogenoalkane

$$-\overset{|}{C}-X \longrightarrow -\overset{|}{C}-NH_2 \quad \text{(where X = Cl, Br or I)}$$

The halogenoalkane is heated under pressure with an excess of alcoholic ammonia. Evaporation of the solvent and ammonia leaves the crude amine which may be contaminated with secondary and tertiary amines. Fractional distillation, under reduced pressure, if necessary, is often used to separate out the primary amine.

$$\bigcirc\!\!\!\!-CH_2Cl \longrightarrow \boxed{\begin{array}{c}NH_3/C_2H_5OH\\ \text{pressure}\end{array}} \longrightarrow \bigcirc\!\!\!\!-CH_2NH_2$$

yield 53%

Reduction of a nitroarene

$$\bigcirc\!\!\!\!-NO_2 \longrightarrow \bigcirc\!\!\!\!-NH_2$$

Iron filings are added to an alcoholic solution of the nitroarene and then concentrated hydrochloric acid is dripped in. The amine so formed can only be extracted into an organic solvent after the solution has been neutralized. This is because amines in acidic conditions are always present as the ionic alkylammonium chloride (amine hydrochloride), $RNH_3^+Cl^-$.

$$\text{C}_6\text{H}_5\text{-NO}_2 \xrightarrow{\begin{array}{c}\text{Fe/HCl,}\\\text{then base}\end{array}} \text{C}_6\text{H}_5\text{-NH}_2$$

yield 86%

Preparation of Nitriles

Substitution of a halogenoalkane

$$-\overset{\shortmid}{\underset{\shortmid}{\text{C}}}-\text{X} \longrightarrow -\overset{\shortmid}{\underset{\shortmid}{\text{C}}}-\text{C}{=}\text{N} \quad \text{(where X = Cl, Br or I)}$$

Heating a primary halogenoalkane under reflux with an ethanolic solution of potassium cyanide will give good yields of the nitrile. Secondary halogenoalkanes give only mediocre yields of nitrile and tertiary halogenoalkanes undergo elimination to form an alkene, rather than substitution to form a nitrile.

$$CH_3CH_2CH_2Br \longrightarrow \boxed{\text{KCN}} \longrightarrow CH_3CH_2CH_2CN$$

Dehydration of an amide

$$-\text{C}\overset{\displaystyle O}{\underset{\displaystyle NH_2}{\Big<}} \longrightarrow -\text{C}{\equiv}\text{N}$$

Phosphorus pentoxide, P_2O_5, is a powerful dehydrating agent and gives good yields of nitrile when stirred with a dry solution of an amide.

$$\begin{array}{c}CH_3\\ \\ CH_3\end{array}\!\!\!\!>\!\!CH_2CONH_2 \longrightarrow \boxed{P_2O_5} \longrightarrow \begin{array}{c}CH_3\\ \\ CH_3\end{array}\!\!\!\!>\!\!CH_2CN$$

Dehydration of an oxime

$$\begin{array}{c}H\\ \\ \end{array}\!\!\!\!>\!\!C{=}N\!\!\underset{\displaystyle OH}{\diagdown} \longrightarrow -\text{C}{\equiv}\text{N}$$

The oxime of an aldehyde can be readily dehydrated. Heating the oxime with ethanoic anhydride, for example, will give good yields of the nitrile.

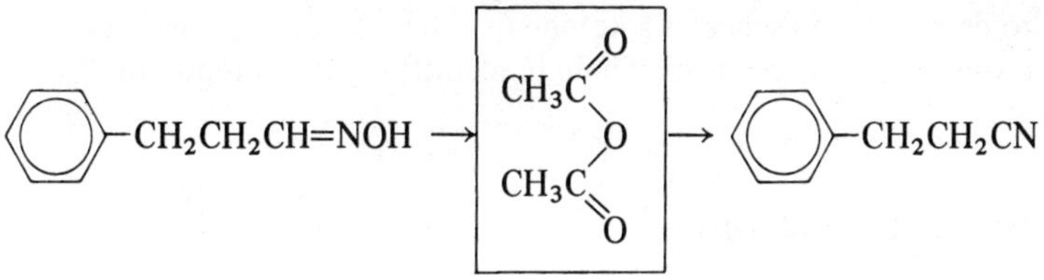

Substitution of an arenediazonium ion

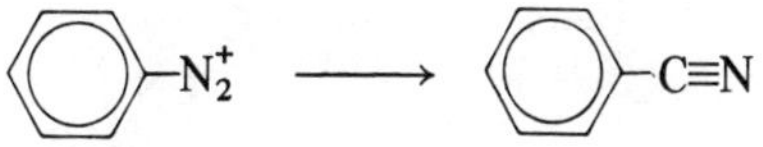

Warming a solution of an arenediazonium salt with copper(I) cyanide, $Cu_2(CN)_2$, will form the aryl cyanide.

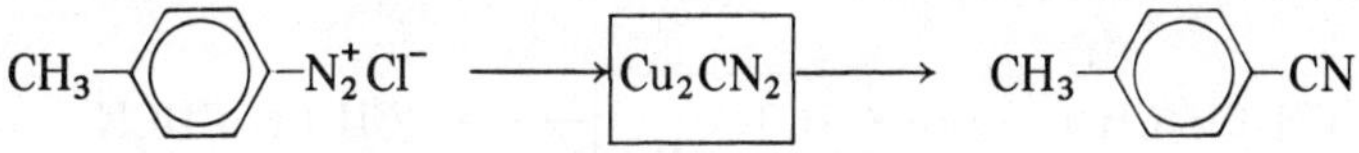

Preparation of Epoxides

Addition to an alkene

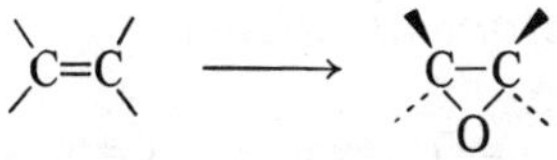

Industrially the epoxides of ethene and propene are prepared in large quantities by oxidation with oxygen and a silver catalyst. A common laboratory procedure is to use an oxoacid, such as peroxoethanoic acid (peracetic acid) to provide the oxygen.

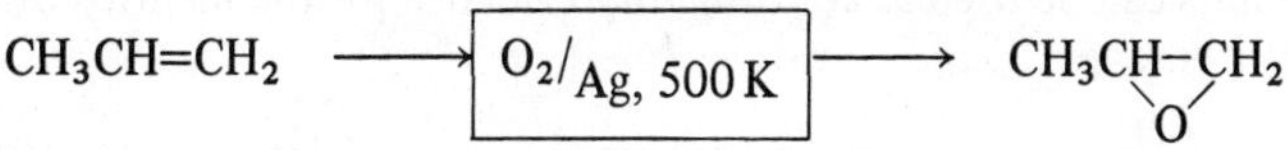

$$CH_3CH=CH_2 \longrightarrow \boxed{O_2/_{Ag,\ 500\,K}} \longrightarrow CH_3CH{-}CH_2$$

Preparation of 2,4-dinitrophenylhydrazones

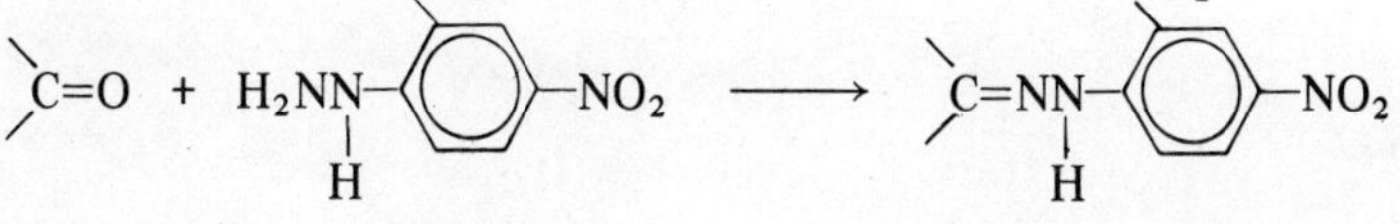

Addition of an aldehyde or ketone to a solution containing 2,4-dinitrophenyl-hydrazine dissolved in a mixture of sulphuric acid and ethanol, gives a red or yellow precipitate of the 2,4-dinitrophenylhydrazone. This reaction may be used as a test to detect the presence of a ketone or aldehyde, or to prepare a derivative whose melting point may help in identifying the compound.

Preparation of Cyanohydrins

Addition to a ketone or aldehyde

$$\text{C=O} \longrightarrow \text{C} \begin{matrix} \text{OH} \\ \text{CN} \end{matrix}$$

Hydrogen cyanide, generated by adding acid, dropwise, to an excess of sodium cyanide, will add to a ketone or aldehyde to form 2-hydroxyalkanonitrile (cyanohydrin). This reaction will only occur in basic conditions, which is why excess sodium cyanide must always be present.

$$CH_3CH_2CHO \longrightarrow \boxed{HCN/_{base}} \longrightarrow CH_3CH_2\underset{OH}{CHCN}$$

Preparation of Oximes

Condensation with a ketone or aldehyde

$$\text{C=O} \longrightarrow \text{C=N}\diagdown_{OH}$$

Warming an aldehyde or ketone with an ethanolic solution of hydroxylammonium chloride (the hydrochloride of hydroxylamine) will produce a high yield of the oxime. Oximes are usually crystalline and are often used as derivatives, whose melting points can be used as confirmatory evidence for the identity of the aldehyde or ketone.

$$\begin{matrix} CH_3 \\ CH_3 \end{matrix} \text{C=O} \longrightarrow \boxed{NH_2OH} \longrightarrow \begin{matrix} CH_3 \\ CH_3 \end{matrix} \text{C=NOH}$$

yield 95%

Two-step Syntheses

Two-step Syntheses of Halides

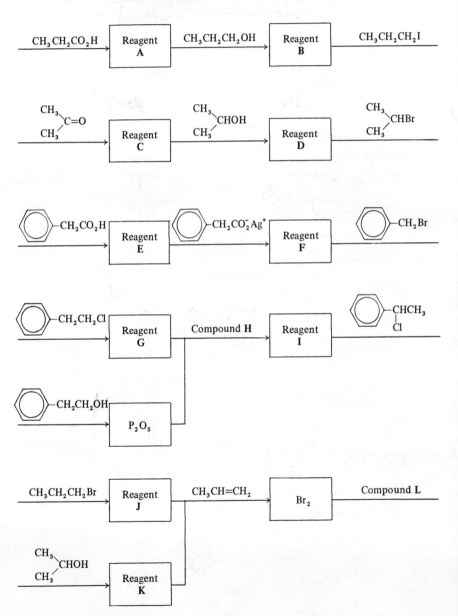

Two-step Syntheses of Carboxylic Acids

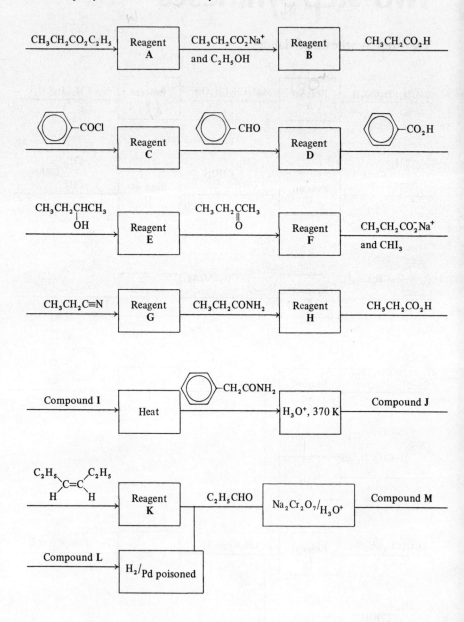

Two-step Syntheses of Alcohols

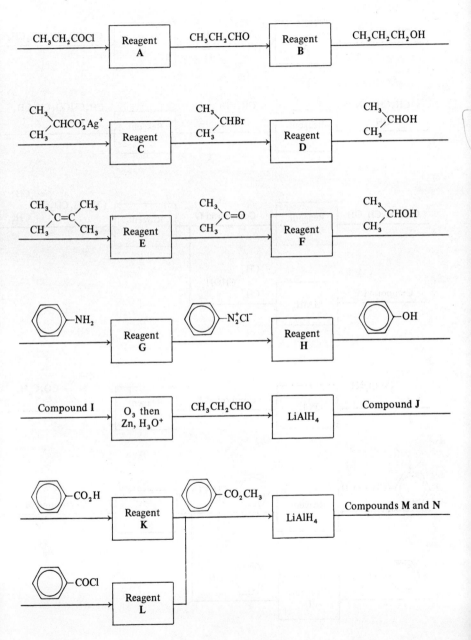

Two-step Syntheses of Esters

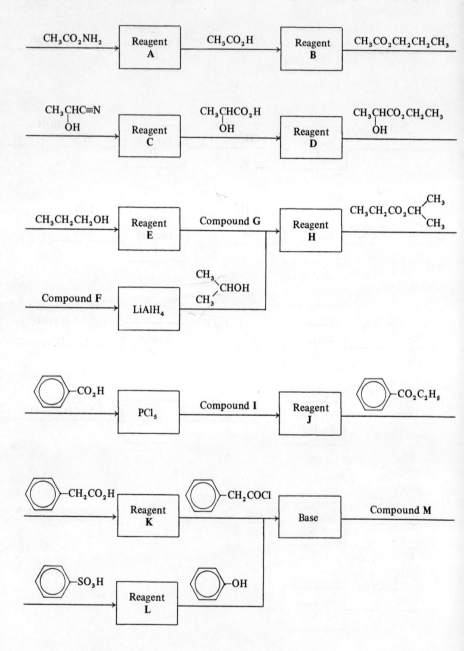

Two-step Syntheses of Aldehydes and Ketones

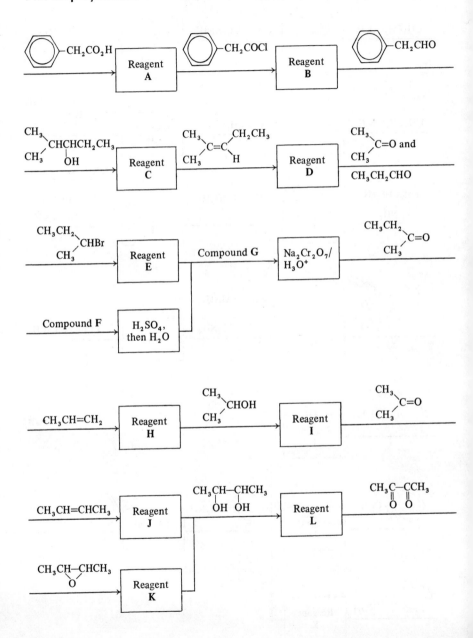

Two-step Syntheses of Amides

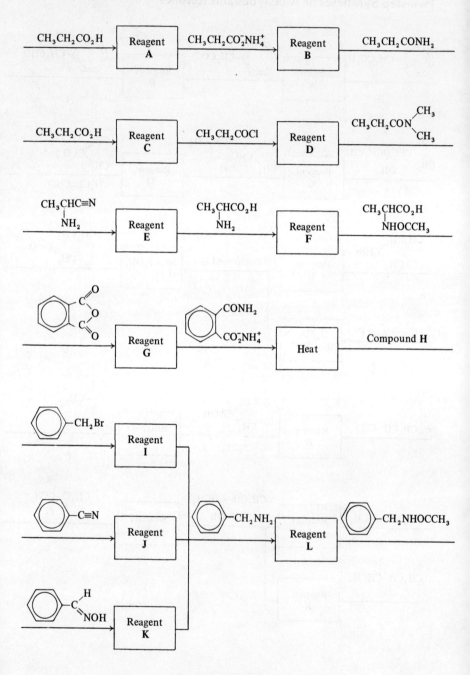

Two-step Syntheses of Amines

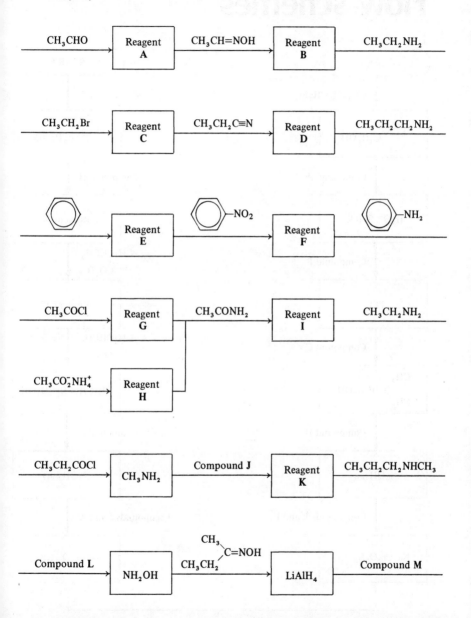

Flow-schemes

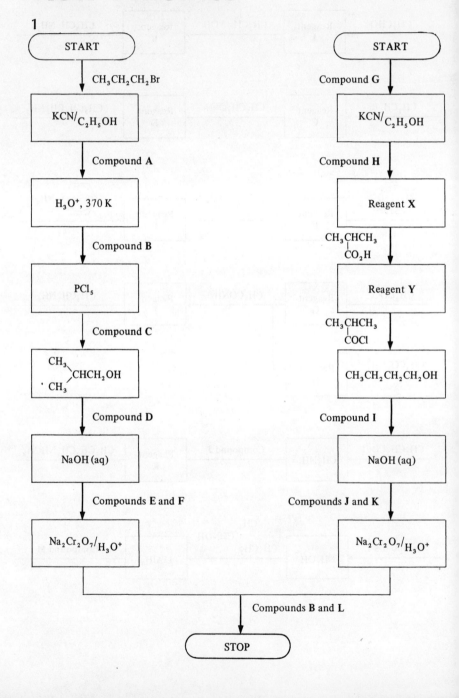

1

START
↓ CH₃CH₂CH₂Br

KCN/C₂H₅OH
↓ Compound A

H₃O⁺, 370 K
↓ Compound B

PCl₅
↓ Compound C

$\begin{matrix} CH_3 \\ CH_3 \end{matrix}$CHCH₂OH
↓ Compound D

NaOH (aq)
↓ Compounds E and F

Na₂Cr₂O₇/H₃O⁺

START
↓ Compound G

KCN/C₂H₅OH
↓ Compound H

Reagent X
↓ CH₃CHCH₃ / CO₂H

Reagent Y
↓ CH₃CHCH₃ / COCl

CH₃CH₂CH₂CH₂OH
↓ Compound I

NaOH (aq)
↓ Compounds J and K

Na₂Cr₂O₇/H₃O⁺

↓ Compounds B and L

STOP

2

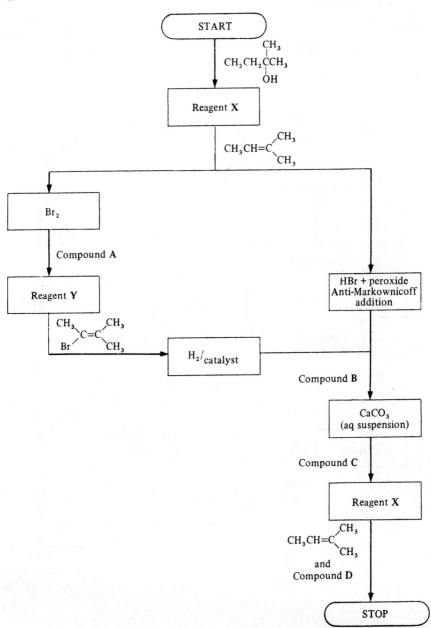

3

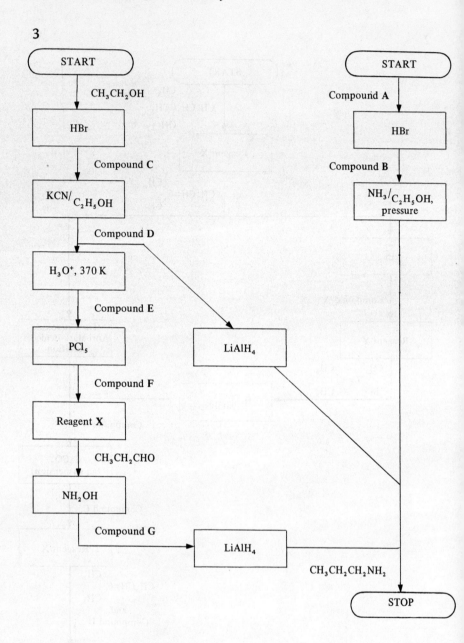

4

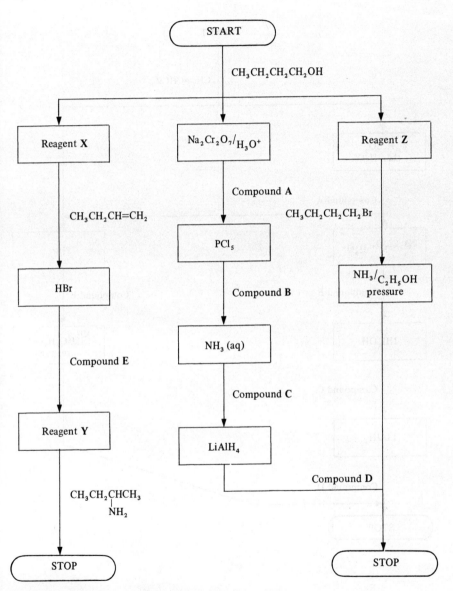

5

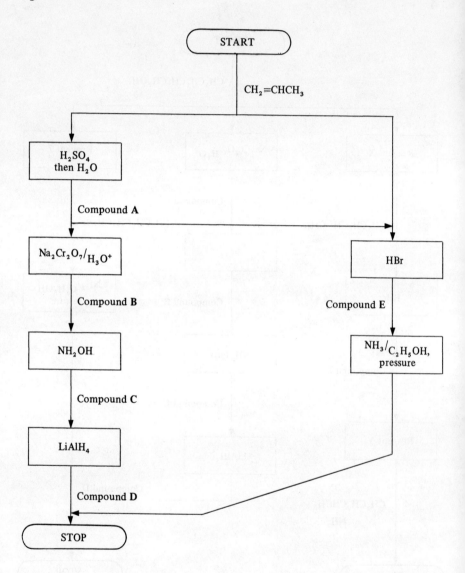

6

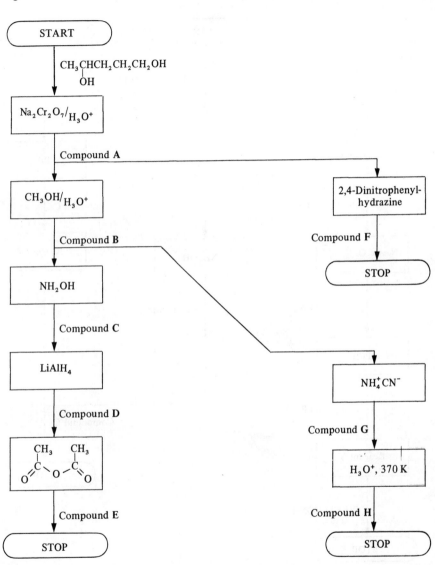

7

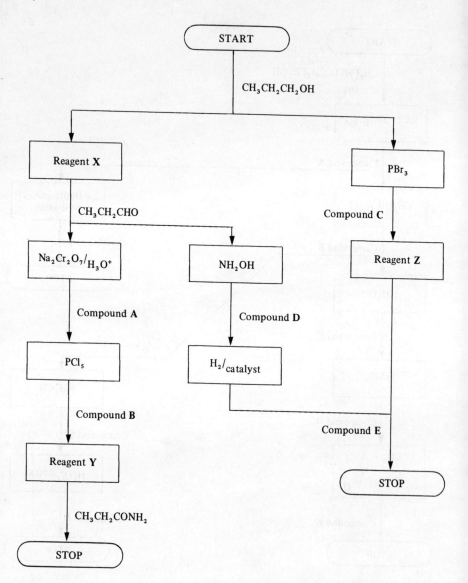

8

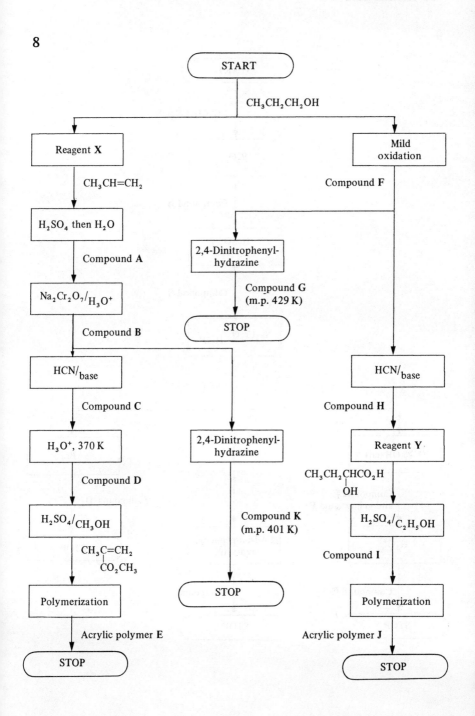

9

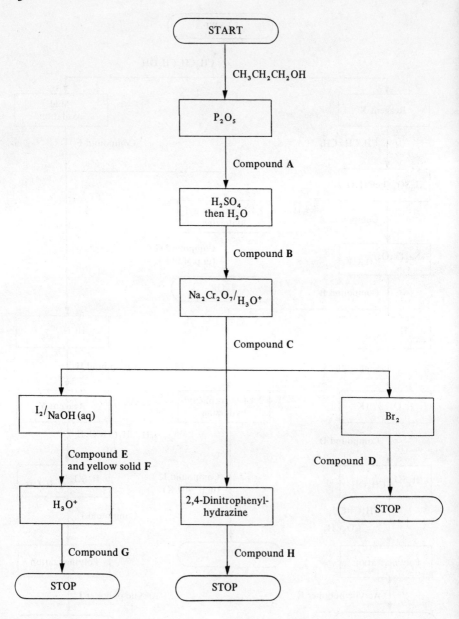

10

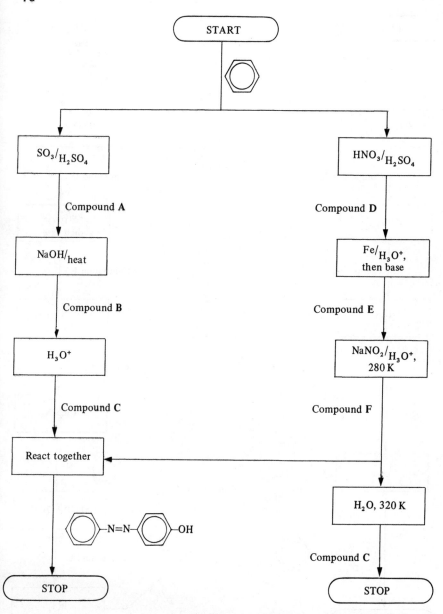

11

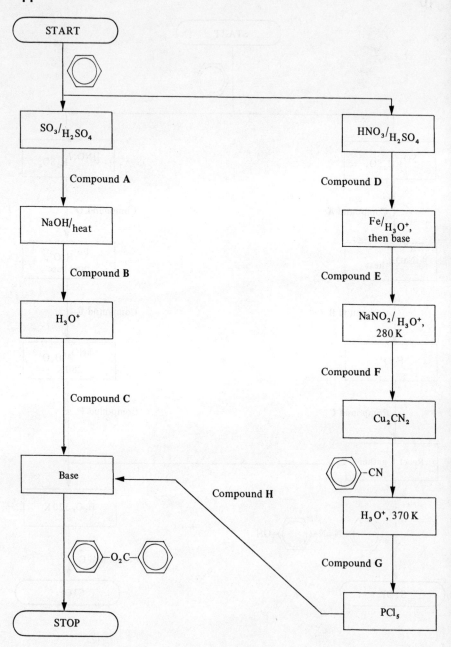

12

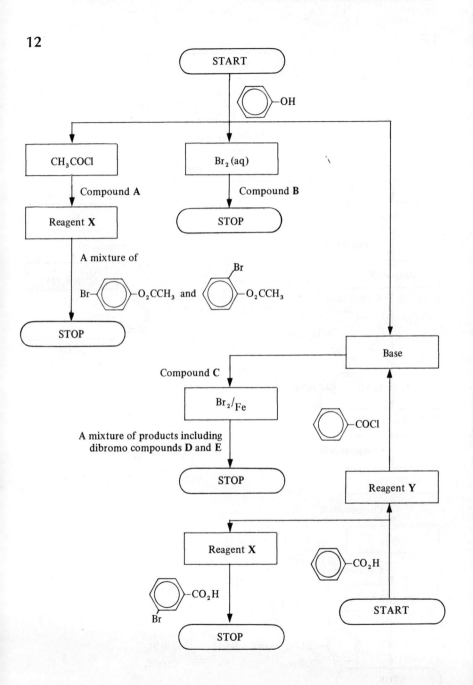

13

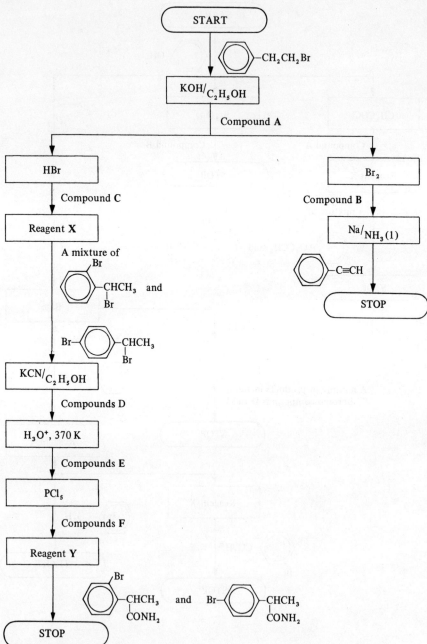

14

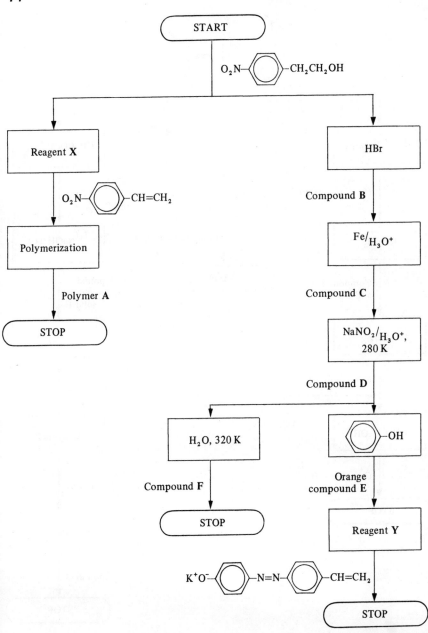

15

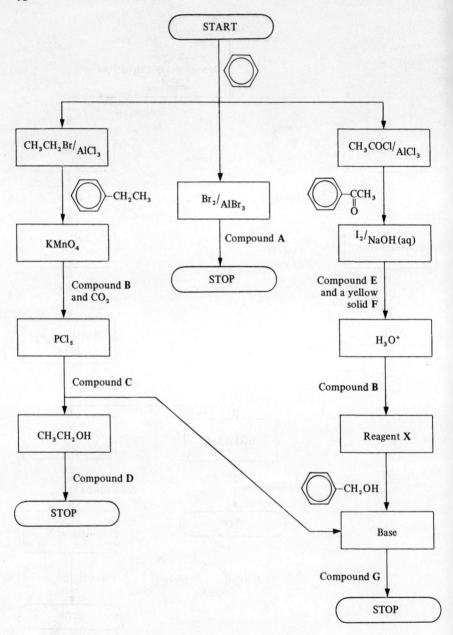

16

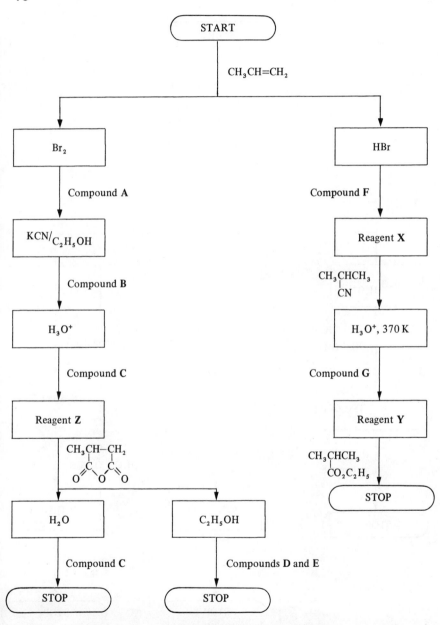

17

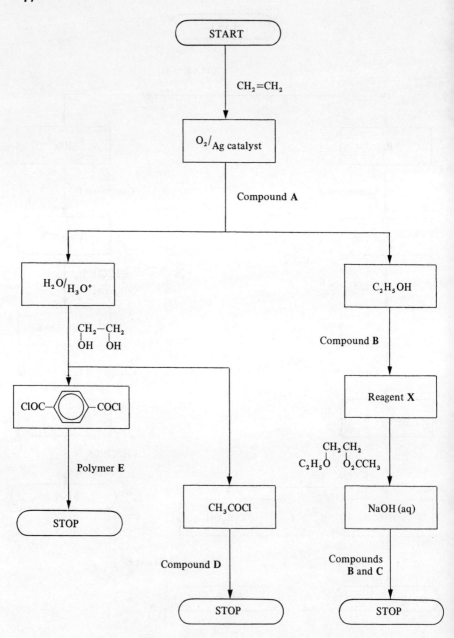

18

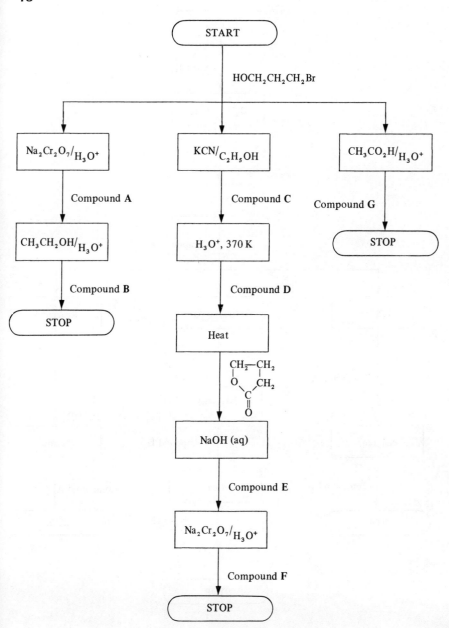

19

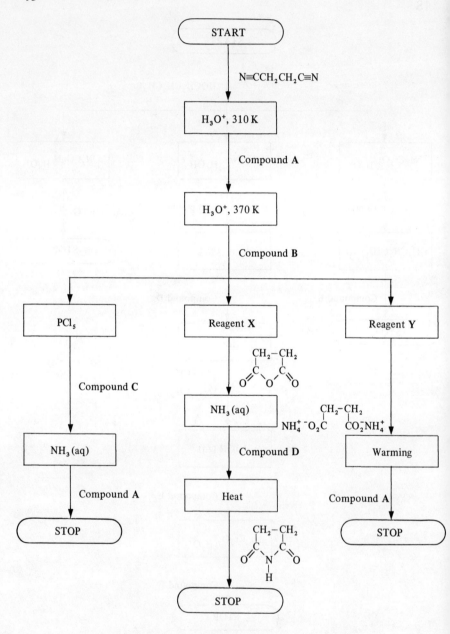

20

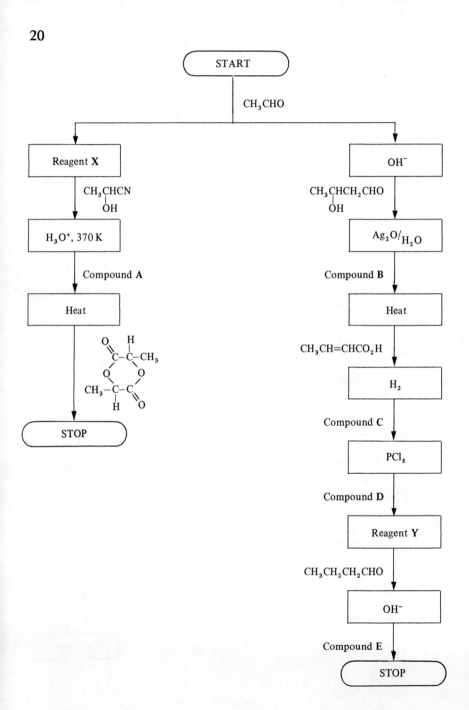

21

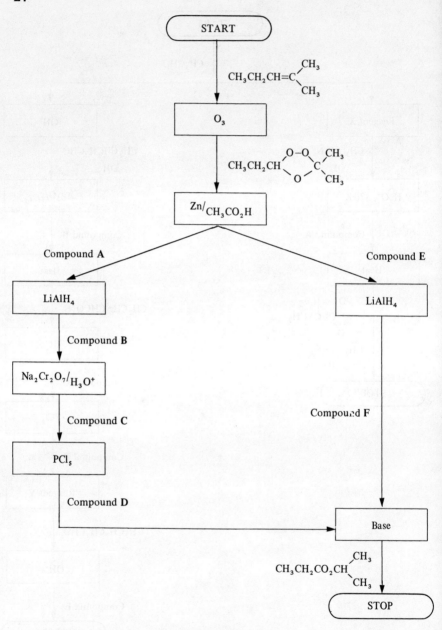

22

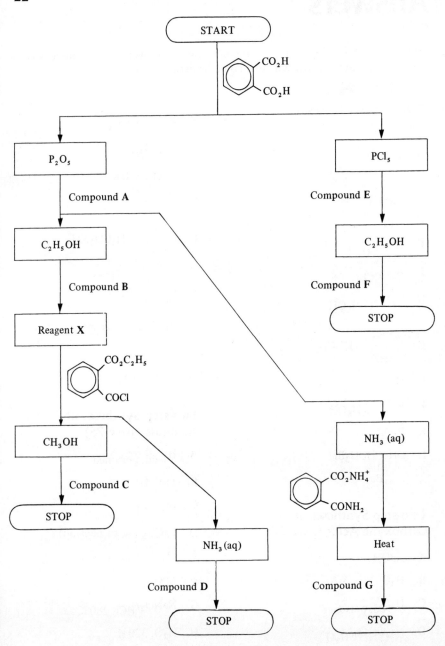

Answers

It should be borne in mind that although only one answer is shown there may often be a number of reagents that could be used.

Two-step Syntheses of Halides (page 23)

A $LiAlH_4$

B PI_3

C $NaBH_4$

D HBr

E Ag_2O/H_2O

F Br_2/CCl_4

G KOH/C_2H_5OH

H —CH=CH$_2$

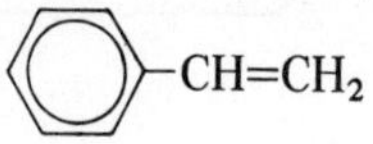

I HCl (g)

J KOH/C_2H_5OH

K P_2O_5

L CH_3CHCH_2Br
       |
       Br

Two-step Syntheses of Carboxylic Acids (page 24)

A $NaOH$

B H_3O^+

C H_2/Pd, poisoned

D $Na_2Cr_2O_7/H_3O^+$

E $Na_2Cr_2O_7/H_3O^+$

F $I_2/NaOH$

G H_3O^+, 320 K

H H_3O^+, 370 K

I —CH$_2$CO$_2^-$NH$_4^+$

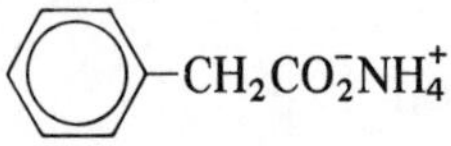

J —CH$_2$CO$_2$H

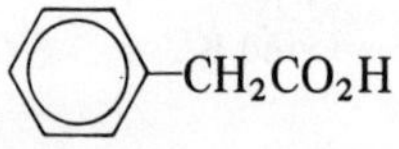

K O_3 then Zn, H_3O^+

L C_2H_5COCl

M $C_2H_5CO_2H$

Two-step Syntheses of Alcohols (page 25)

A H_2/Pd, poisoned

B $LiAlH_4$

C Br_2/CCl_4

D $CaCO_3$ (aq suspension)

E O_3 then Zn, H_3O^+

F $LiAlH_4$

G $NaNO_2/HCl$, 280 K

H H_2O, 320 K

I CH_3CH_2 $\diagdown$ $C=C$ $\diagup$ CH_2CH_3 with H below left and H below right

$$CH_3CH_2\underset{H}{\overset{}{C}}=\underset{H}{\overset{}{C}}CH_2CH_3$$

J $CH_3CH_2CH_2OH$

K $CH_3OH/_{H_2SO_4}$

L CH_3OH

M   ⬡—CH_2OH

N CH_3OH

Two-step Syntheses of Esters (page 26)

A H_3O^+, 370 K

B $CH_3CH_2CH_2OH/_{H_2SO_4}$

C H_3O^+, 370 K

D $CH_3CH_2OH/_{H_2SO_4}$

E $Na_2Cr_2O_7/_{H_3O^+}$

F CH_3CCH_3 with $\overset{\|}{O}$

$$CH_3\underset{\underset{O}{\|}}{C}CH_3$$

G $CH_3CH_2CO_2H$

H H_2SO_4

I   ⬡—$COCl$

J $C_2H_5OH/_{base}$

K PCl_5

L $NaOH/_{heat}$, then H_3O^+

M   ⬡—CH_2CO_2—⬡

Two-step Syntheses of Aldehydes and Ketones (page 27)

A PCl_5

B $H_2/_{Pd}$, poisoned

C H_3PO_4

D O_3 then Zn, H_3O^+

E $CaCO_3$ (aq suspension)

F $CH_3CH=CHCH_3$

G $CH_3CH_2CHCH_3$ with OH below

$$CH_3CH_2\underset{OH}{\overset{}{C}}HCH_3$$

H H_2SO_4, then H_2O

I $Na_2Cr_2O_7/_{H_3O^+}$

J $KMnO_4$, (neutral)

K $H_2O/_{H_3O^+}$

L $Na_2Cr_2O_7/_{H_3O^+}$

Two-step Syntheses of Amides (page 28)

A NH_3 (aq)

B   Heat

C PCl_5

D CH_3 $\diagdown$ NH / CH_3 $\diagup$

E H_3O^+, 370 K

F CH_3COCl

G NH_3 (aq)

H

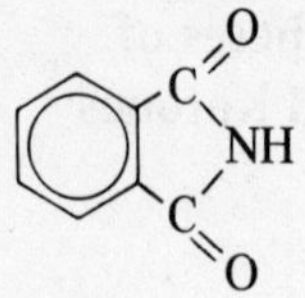

I NH_3/C_2H_5OH, pressure

J $LiAlH_4$

K $LiAlH_4$

L CH_3COCl

Two-step Syntheses ᶠ Amines (page 29)

A $NH_2OH.HCl$

B $LiAlH_4$

C KCN

D $LiAlH_4$

E HNO_3/H_2SO_4

F Fe/H_3O^+

G $NH_3(aq)$

H Heat

I $LiAlH_4$

J $CH_3CH_2CONHCH_3$

K $LiAlH_4$

L $CH_3CH_2CCH_3$
   $\overset{\|}{O}$

M $\underset{CH_3CH_2}{\overset{CH_3}{>}}CHNH_2$

Flow-scheme 1 (page 30)

A $CH_3CH_2CH_2CN$

B $CH_3CH_2CH_2CO_2H$

C $CH_3CH_2CH_2COCl$

D $CH_3CH_2CH_2CO_2CH_2CH\overset{CH_3}{\underset{CH_3}{<}}$

E $CH_3CH_2CH_2CO_2^- Na^+$

F $\underset{CH_3}{\overset{CH_3}{>}}CHCH_2OH$

G CH_3CHCH_3
   $\overset{|}{Br}$

H CH_3CHCH_3
   $\overset{|}{CN}$

I $\underset{CH_3}{\overset{CH_3}{>}}CHCO_2CH_2CH_2CH_2CH_3$

J $\underset{CH_3}{\overset{CH_3}{>}}CHCO_2^-Na^+$

K $CH_3CH_2CH_2CH_2OH$

L $\underset{CH_3}{\overset{CH_3}{>}}CHCO_2H$

X H_3O^+, 370 K

Y PCl_5

Flow-scheme 2 (page 31)

A $CH_3\overset{CH_3}{\underset{|}{C}}HCCH_3$
   $\underset{Br\ Br}{}$

B $CH_3CHCHCH_3$
   $\underset{Br\ CH_3}{}$

C $CH_3CHCHCH_3$
$\quad\quad\;\; |\;\;\;\; |$
$\quad\quad\; HO\;\; CH_3$

D $CH_2{=}CHCHCH_3$
$\quad\quad\quad\quad\quad |$
$\quad\quad\quad\quad\; CH_3$

X P_2O_5

Y KOH/C_2H_5OH

Flow-scheme 3 (page 32)

A $CH_3CH_2CH_2OH$

B $CH_3CH_2CH_2Br$

C CH_3CH_2Br

D CH_3CH_2CN

E $CH_3CH_2CO_2H$

F CH_3CH_2COCl

G $CH_3CH_2CH{=}NOH$

X $H_2/Pd,\; poisoned$

Flow-scheme 4 (page 33)

A $CH_3CH_2CH_2CO_2H$

B $CH_3CH_2CH_2COCl$

C $CH_3CH_2CH_2CONH_2$

D $CH_3CH_2CH_2CH_2NH_2$

E $CH_3CH_2CHCH_3$
$\quad\quad\quad\quad\; |$
$\quad\quad\quad\quad Br$

X P_2O_5

Y $NH_3/C_2H_5OH,\; pressure$

Z HBr

Flow-scheme 5 (page 34)

A CH_3CHCH_3
$\quad\quad\; |$
$\quad\quad OH$

B CH_3CCH_3
$\quad\quad\;\, \|$
$\quad\quad\;\, O$

C CH_3CCH_3
$\quad\quad\;\, \|$
$\quad\quad NOH$

D CH_3CHCH_3
$\quad\quad\; |$
$\quad\quad NH_2$

E CH_3CHCH_3
$\quad\quad\; |$
$\quad\quad Br$

Flow-scheme 6 (page 35)

A $CH_3CCH_2CH_2CO_2H$
$\quad\quad\;\, \|$
$\quad\quad\;\, O$

B $CH_3CCH_2CH_2CO_2CH_3$
$\quad\quad\;\, \|$
$\quad\quad\;\, O$

C $CH_3CCH_2CH_2CO_2CH_3$
$\quad\quad\;\, \|$
$\quad\quad NOH$

D $CH_3CHCH_2CH_2CH_2OH$
$\quad\quad\;\; |$
$\quad\quad\; NH_2$

E $CH_3CHCH_2CH_2CH_2O_2CCH_3$
$\quad\quad\;\; |$
$\quad\quad NHOCCH_3$

F $CH_3CCH_2CH_2CO_2H$
$\quad\quad\; \|$
$\quad\quad\; N$
$\quad\quad\; |$
$\quad\quad NH{-}\!\!\bigcirc\!\!{-}NO_2$
$\quad\quad\quad\quad\quad |$
$\quad\quad\quad\quad\; NO_2$

G $CH_3\overset{\displaystyle NH_2}{\underset{\displaystyle CN}{\overset{|}{\underset{|}{C}}}}CH_2CH_2CO_2CH_3$

H $CH_3\overset{\displaystyle NH_2}{\underset{\displaystyle CO_2H}{\overset{|}{\underset{|}{C}}}}CH_2CH_2CO_2H$

Flow-scheme 7 (page 36)

A $CH_3CH_2CO_2H$

B CH_3CH_2COCl

C $CH_3CH_2CH_2Br$

D $CH_3CH_2CH=NOH$

E $CH_3CH_2CH_2NH_2$

X $Na_2Cr_2O_7/H_3O^+$, and distil out the product as it forms

Y $NH_3(aq)$

Z NH_3/C_2H_5OH, pressure

Flow-scheme 8 (page 37)

A $CH_3\underset{\displaystyle OH}{\overset{|}{CH}}CH_3$

B $CH_3\overset{\displaystyle \|}{\underset{\displaystyle O}{C}}CH_3$

C $CH_3\overset{\displaystyle OH}{\underset{\displaystyle CN}{\overset{|}{\underset{|}{C}}}}CH_3$

D $CH_3\overset{\displaystyle OH}{\underset{\displaystyle CO_2H}{\overset{|}{\underset{|}{C}}}}CH_3$

E $\left[\!\!\begin{array}{c} CH_3 \\ | \\ -C-CH_2- \\ | \\ CO_2CH_3 \end{array}\!\!\right]_n$

F CH_3CH_2CHO

G $CH_3CH_2CH=NNH-\!\!\!\underset{\displaystyle NO_2}{\bigcirc}\!\!\!-NO_2$

H $CH_3CH_2\underset{\displaystyle OH}{\overset{|}{CH}}CN$

I $CH_3CH=CHCO_2C_2H_5$

J $\left[\!\!\begin{array}{c} CH_3 \\ | \\ -CH-CH- \\ | \\ CO_2C_2H_5 \end{array}\!\!\right]_n$

K $\overset{\displaystyle CH_3}{\underset{\displaystyle CH_3}{>}}C=NNH-\!\!\!\underset{\displaystyle NO_2}{\bigcirc}\!\!\!-NO_2$

X P_2O_5

Y H_3O^+, 370 K

Flow-scheme 9 (page 38)

A $CH_3CH=CH_2$

B $CH_3\underset{\displaystyle OH}{\overset{|}{CH}}CH_3$

C $CH_3\overset{\displaystyle \|}{\underset{\displaystyle O}{C}}CH_3$

D $CH_3\overset{\displaystyle \|}{\underset{\displaystyle O}{C}}CH_2Br$

E $CH_3CO_2^-Na^+$

F CHI_3

G CH_3CO_2H

H (CH₃)₂C=NNH–C₆H₃(NO₂)₂ structure

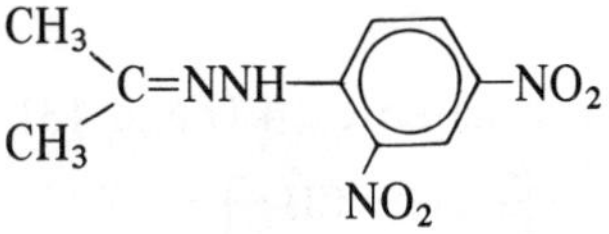

Flow-scheme 10 (page 39)

A —SO_3H

B —O^-Na^+

C —OH

D —NO_2

E —NH_2

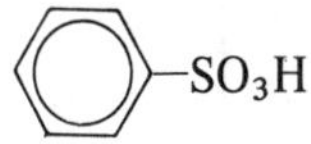

F —N_2^+

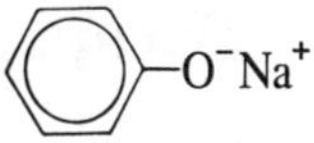

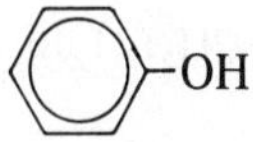

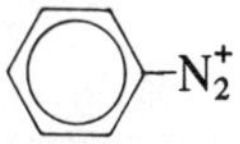

Flow-scheme 11 (page 40)

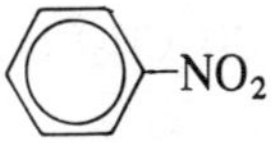

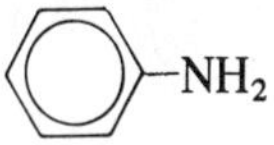

A —SO_3H

B —O^-Na^+

C —OH

D —NO_2

E —NH_2

F —N_2^+

G —CO_2H

H —COCl

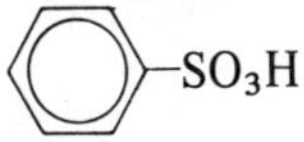

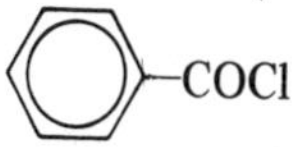

Flow-scheme 12 (page 41)

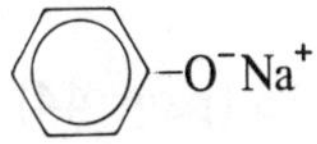

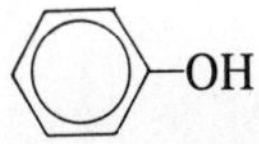

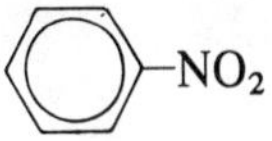

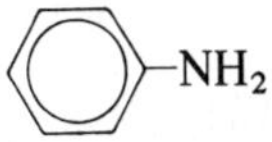

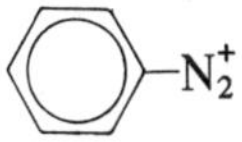

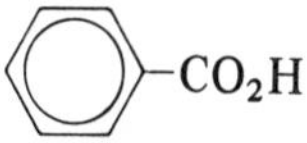

A —O_2CCH_3

B Br——OH (with Br substituents)

C —CO_2—

D —CO_2——Br (with Br)

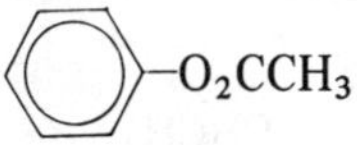

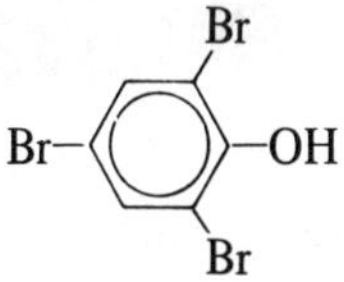

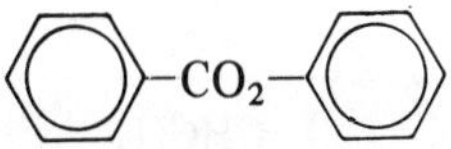

E —CO_2—(with Br)

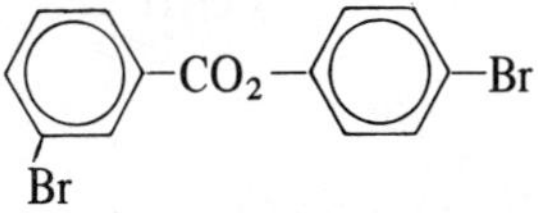

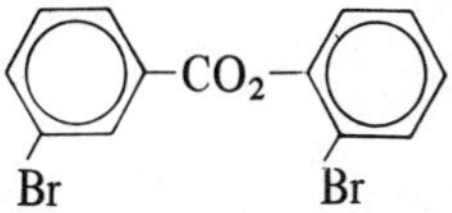

X Br_2/Fe

Y PCl_5

Flow-scheme 13 (page 42)

A C₆H₅—CH=CH₂

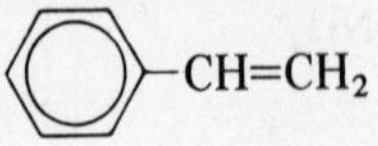

B C₆H₅—CHCH₂Br
          |
          Br

C C₆H₅—CHCH₃
          |
          Br

D
```
    Br
     |
  [C₆H₄]—CHCH₃   and   Br—[C₆H₄]—CHCH₃
              |                        |
              CN                       CN
```

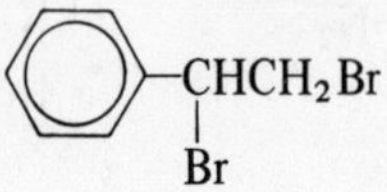

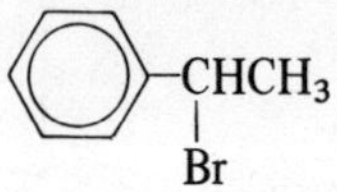

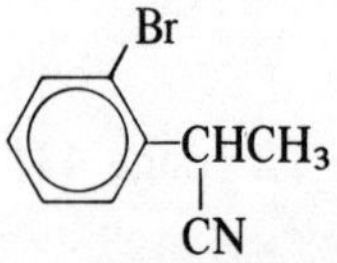

E
```
    Br
     |
  [C₆H₄]—CHCH₃   and   Br—[C₆H₄]—CHCH₃
              |                        |
              CO₂H                     CO₂H
```

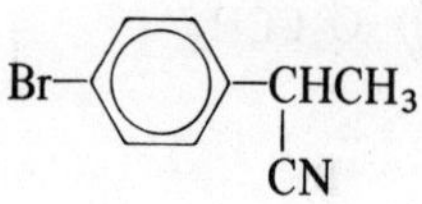

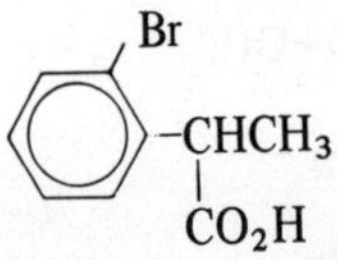

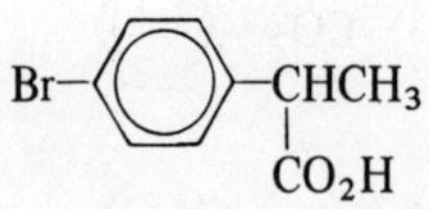

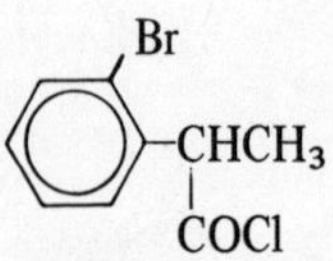

F
```
    Br
     |
  [C₆H₄]—CHCH₃   and   Br—[C₆H₄]—CHCH₃
              |                        |
              COCl                     COCl
```

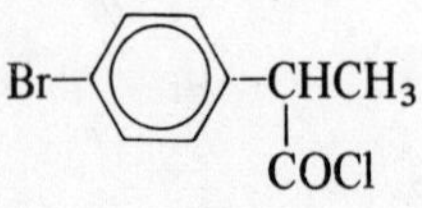

X Br_2/Fe

Y NH_3 (aq)

Flow-scheme 14 (page 43)

A
$$\left[\; CH-CH_2\; \right]_n$$
with phenyl ring bearing NO_2

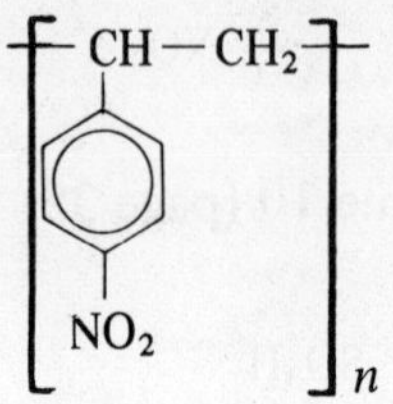

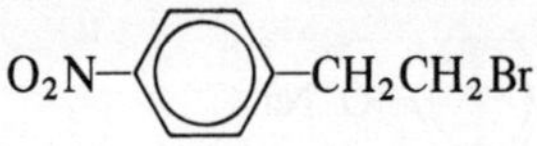

B O_2N—[C₆H₄]—CH_2CH_2Br

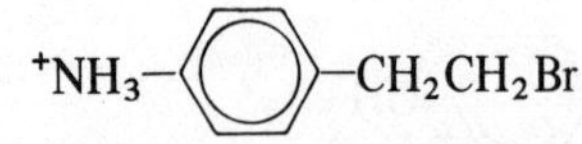

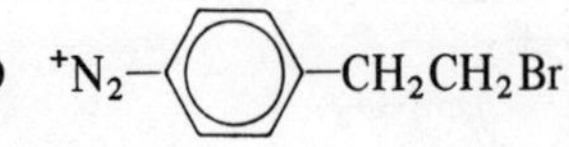

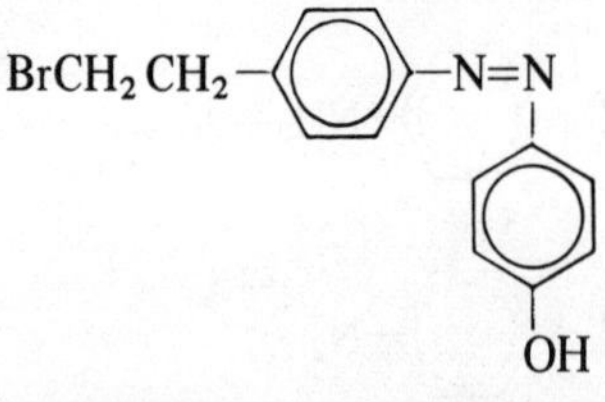

C $^{+}NH_3$—[C₆H₄]—CH_2CH_2Br

D $^{+}N_2$—[C₆H₄]—CH_2CH_2Br

E $BrCH_2CH_2$—[C₆H₄]—N=N—[C₆H₄]—OH

F HO—[C₆H₄]—CH_2CH_2Br

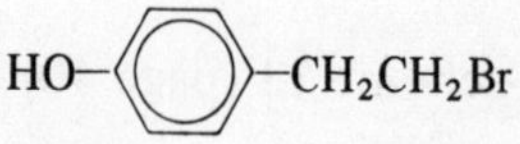

X H_3PO_4

Y KOH/C_2H_5OH(aq)

Flow-scheme 15 (page 44)

A C₆H₅—Br

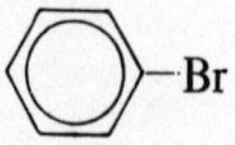

B C₆H₅—CO₂H

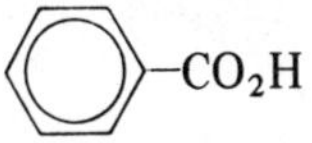

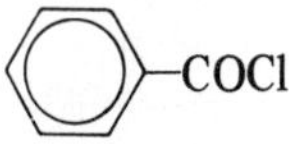

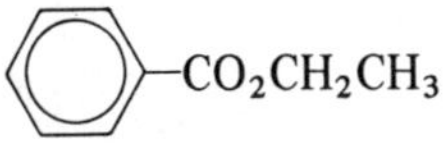

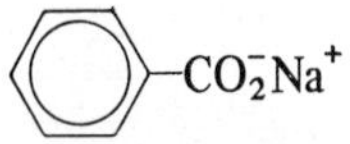

C C₆H₅—COCl

D C₆H₅—CO₂CH₂CH₃

E C₆H₅—CO₂⁻Na⁺

F CHI₃

G C₆H₅—CO₂CH₂—C₆H₅

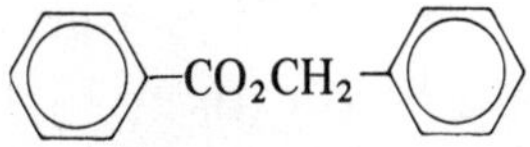

X LiAlH₄

Flow-scheme 16 (page 45)

A CH₃CHCH₂Br
    |
    Br

B CH₃CHCH₂CN
    |
    CN

C CH₃CHCH₂CO₂H
    |
    CO₂H

D CH₃CHCH₂CO₂H
    |
    CO₂C₂H₅

E CH₃CHCH₂CO₂C₂H₅
    |
    CO₂H

F CH₃CHCH₃
    |
    Br

G CH₃CHCH₃
    |
    CO₂H

X KCN/C₂H₅OH

Y C₂H₅OH/H₃O⁺

Z Heat

Flow-scheme 17 (page 46)

A $\overset{CH_2CH_2}{\underset{O}{\diagdown\diagup}}$

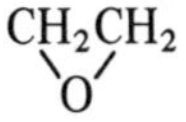

B C₂H₅OCH₂CH₂OH

C CH₃CO₂⁻Na⁺

D CH₃CO₂CH₂CH₂O₂CCH₃

E [—OCH₂CH₂O₂C—C₆H₄—CO—]ₙ

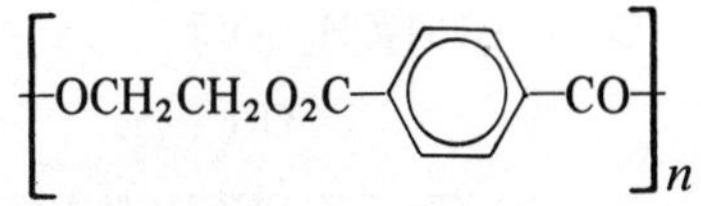

X $\overset{CH_3}{\underset{O}{\diagdown C}}\overset{}{\underset{O}{}}\overset{CH_3}{\underset{O}{C\diagup}}$

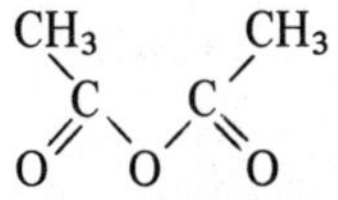

Flow-scheme 18 (page 47)

A HO₂CCH₂CH₂Br

B CH₃CH₂O₂CCH₂CH₂Br

C HOCH₂CH₂CH₂CN

D HOCH₂CH₂CH₂CO₂H

E HOCH₂CH₂CH₂CO₂⁻Na⁺

F HO₂CCH₂CH₂CO₂H

G CH₃CO₂CH₂CH₂CH₂Br

Flow-scheme 19 (page 48)

A H₂NOCCH₂CH₂CONH₂

B HO₂CCH₂CH₂CO₂H

C ClOCCH₂CH₂COCl

D $H_2NOCCH_2CH_2CO_2^-NH_4^+$

X Heat

Y $NH_3(aq)$

Flow-scheme 20 (page 49)

A CH₃CHCO₂H
     |
    OH

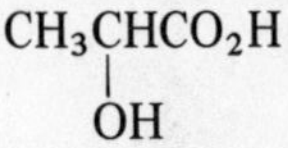

A $\underset{\overset{|}{\underset{OH}{}}}{CH_3CH}CO_2H$

B $\underset{\overset{|}{\underset{OH}{}}}{CH_3CH}CH_2CO_2H$

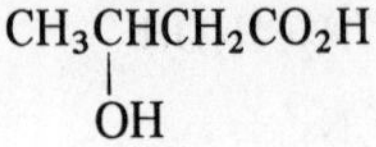

C $CH_3CH_2CH_2CO_2H$

D $CH_3CH_2CH_2COCl$

E $CH_3CH_2CH_2\underset{\overset{|}{\underset{CHO}{}}}{CH}\underset{\overset{|}{\underset{OH}{}}}{CH}CH_2CH_3$

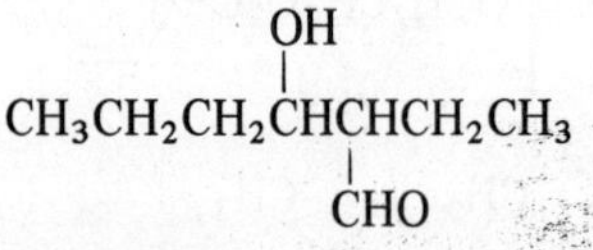

X $HCN/_{base}$

Y $H_2/_{Pd, \text{ poisoned}}$

Flow-scheme 21 (page 50)

A CH_3CH_2CHO

B $CH_3CH_2CH_2OH$

C $CH_3CH_2CO_2H$

D CH_3CH_2COCl

E $\underset{CH_3}{\overset{CH_3}{>}}C=O$

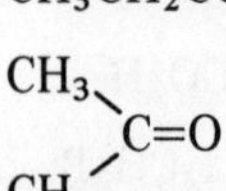

F $\underset{CH_3}{\overset{CH_3}{>}}CHOH$

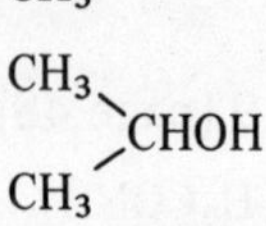

Flow-scheme 22 (page 51)

A

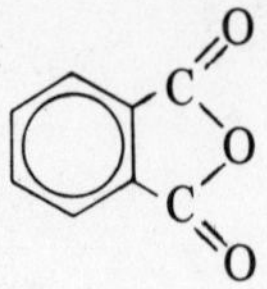

B

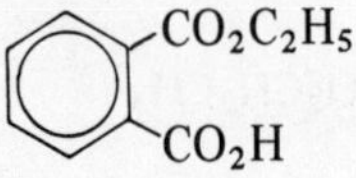

C

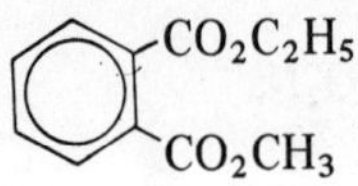

D

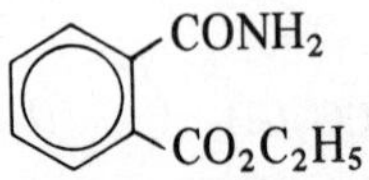

E

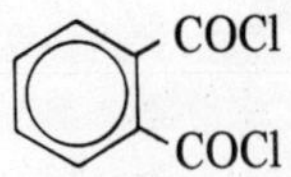

F

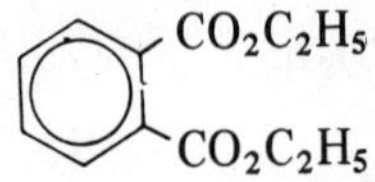

G

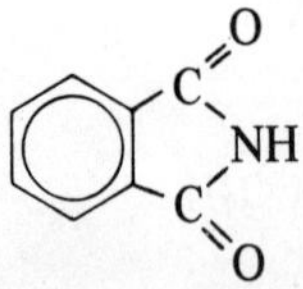

X PCl_5